Fundamentals of Crew Leadership

46101-11

National Center for Construction Education and Research

President: Don Whyte
Director of Product Development: Daniele Stacey
Fundamentals of Crew Leadership Project Manager: Patty Bird
Production Manager: Tim Davis
Quality Assurance Coordinator: Debie Ness
Editor: Chris Wilson
Desktop Publishing Coordinator: James McKay
Production Assistant: Laura Wright

Editorial and production services provided by Topaz Publications, Liverpool, NY
Lead Writer/Project Manager: Tom Burke
Desktop Publisher: Joanne Hart
Art Director: Megan Paye
Permissions Editors: Andrea LaBarge, Alison Richmond

V.1 1/11

Prentice Hall
is an imprint of

www.pearsonhighered.com

ISBN 13: 978-0-13-610652-4

FOREWORD

Work gets done most efficiently if workers are divided into crews with a common purpose. When a crew is formed to tackle a particular job, one person is appointed the leader. This person is usually an experienced craftworker who has demonstrated leadership qualities. To become an effective leader, it helps if you have natural leadership qualities, but there are specific job skills that you must learn in order to do the job well.

This module will teach you the skills you need to be an effective leader, including the ability to communicate effectively; provide direction to your crew; and effectively plan and schedule the work of your crew.

As a crew member, you weren't required to think much about project cost. However, as a crew leader, you need to understand how to manage materials, equipment, and labor in order to work in a cost-effective manner. You will also begin to view safety from a different perspective. The crew leader takes on the responsibility for the safety of the crew, making sure that workers follow company safety polices and have the latest information on job safety issues.

As a crew leader, you become part of the chain of command in your company, the link between your crew and those who supervise and manage projects. As such, you need to know how the company is organized and how you fit into the organization. You will also focus more on company policies than a crew member, because it is up to you to enforce them within your crew. You will represent your team at daily project briefings and then communicate relevant information to your crew. This means learning how to be an effective listener and an effective communicator.

Whether you are currently a crew leader wanting to learn more about the requirements, or a crew member preparing to move up the ladder, this module will help you reach your goal.

This program consists of an Annotated Instructor's Guide (AIG) and a Participant's Manual. The AIG contains a breakdown of the information provided in the Participant's Manual as well as the actual text that the participant will use. The Participant's Manual contains the material that the participant will study, along with self-check exercises and activities, to help in evaluating whether the participant has mastered the knowledge needed to become an effective crew leader.

For the participant to gain the most from this program, it is recommended that the material be presented in a formal classroom setting, using a trained and experienced instructor. If the student is so motivated, he or she can study the material on a self-learning basis by using the material in both the Participant's Manual and the AIG. Recognition through the National Registry is available for the participants provided the program is delivered through an Accredited Sponsor by a Master Trainer or ICTP instructor. More details on this program can be received by contacting the National Center for Construction Education and Research at www.nccer.org.

Participants in this program should note that some examples provided to reinforce the material may not apply to the participant's exact work, although the process will. Every company has its own mode of operation. Therefore, some topics may not apply to every participant's company. Such topics have been included because they are important considerations for prospective crew leaders throughout the industries supported by NCCER.

A Note to NCCER Instructors and Trainees

If you are training through an Accredited NCCER Sponsor company, note that you may be eligible for dual credentials upon completion. When submitting Form 200, indicate completion of the two module numbers that apply to *Fundamentals of Crew Leadership* – 46101-11 (from NCCER's Contren® Management Series) or the applicable module in a craft training program, such as Module 26413 from NCCER's Electrical curriculum. Transcripts will be issued accordingly.

Contents ──────────

Topics to be presented in this module include:

Contents (continued) ━━━━━━━━━━━━

Contents (continued)

Contents (continued)

Figures and Tables

Acknowledgments

This curriculum was revised as a result of the farsightedness
and leadership of the following sponsors:

ABC South Texas Chapter
HB Training & Consulting
Turner Industries Group, LLC

University of Georgia
Vision Quest Academy
Willmar Electric Service

This curriculum would not exist were it not for the dedication and unselfish energy of
those volunteers who served on the Authoring Team. A sincere thanks is extended to the following:

John Ambrosia
Harold (Hal) Heintz
Mark Hornbuckle
Jonathan Liston

Jay Tornquist
Wayne Tyson
Antonio "Tony" Vazquez

NCCER Partners

American Fire Sprinkler Association

Associated Builders and Contractors, Inc.

Associated General Contractors of America

Association for Career and Technical Education

Association for Skilled and Technical Sciences

Carolinas AGC, Inc.

Carolinas Electrical Contractors Association

Center for the Improvement of Construction
Management and Processes

Construction Industry Institute

Construction Users Roundtable

Design Build Institute of America

Merit Contractors Association of Canada

Metal Building Manufacturers Association

NACE International

National Association of Manufacturers

National Association of Minority Contractors

National Association of Women in Construction

National Insulation Association

National Ready Mixed Concrete Association

National Technical Honor Society

National Utility Contractors Association

NAWIC Education Foundation

North American Technician Excellence

Painting & Decorating Contractors of America

Portland Cement Association

SkillsUSA

Steel Erectors Association of America

U.S. Army Corps of Engineers

University of Florida, M.E. Rinker School of
Building Construction

Women Construction Owners & Executives, USA

Objectives

Upon completion of this section, you should be able to:

1. Describe the opportunities in the construction and power industries.
2. Describe how workers' values change over time.
3. Explain the importance of training and safety for the leaders in the construction and power industries.
4. Describe how new technologies are beneficial to the construction and power industries.
5. Identify the gender and minority issues associated with a changing workforce.
6. Describe what employers can do to prevent workplace discrimination.
7. Differentiate between formal and informal organizations.
8. Describe the difference between authority, responsibility, and accountability.
9. Explain the purpose of job descriptions and what they should include.
10. Distinguish between company policies and procedures.

1.0.0 INDUSTRY TODAY

Today's managers, supervisors, and crew leaders face challenges different from those of previous generations of leaders. To be a leader in industry today, it is essential to be well prepared. Today's crew leaders must understand how to use various types of new technology. In addition, they must have the knowledge and skills needed to manage, train, and communicate with a culturally diverse workforce whose attitudes toward work may differ from those of earlier generations and cultures. These needs are driven by changes in the workforce itself and in the work environment, and include the following:

- A shrinking workforce
- The growth of technology
- Changes in employee attitudes and values
- The emphasis on bringing women and minorities into the workforce
- The growing number of foreign-born workers
- Increased emphasis on workplace health and safety
- Greater focus on education and training

1.1.0 The Need for Training

Effective craft training programs are necessary if the industry is to meet the forecasted worker demands. Many of the skilled, knowledgeable craftworkers, crew leaders, and managers—the so-called baby boomers—have reached retirement age. In 2010, these workers who were born between 1946 and 1964, represented 38 percent of the workforce. Their departure leaves a huge vacuum across the industry spectrum. The Department of Labor (DOL) concludes that the best way for industry to reduce shortages of skilled workers is to create more education and training opportunities. The DOL suggests that companies and community groups form partnerships and create apprenticeship programs. Such programs could provide younger workers, including women and minorities, with the opportunity to develop job skills by giving them hands-on experience.

When training workers, it is important to understand that people learn in different ways. Some people learn by doing, some people learn by watching or reading, and others need step-by-step instructions as they are shown the process. Most people learn best by a combination of styles. It is important to understand what kind of a learner you are teaching, because if you learn one way, you tend to teach the way you learn. Have you ever tried to teach somebody and failed, and then another person successfully teaches the same thing in a different way? A person who acts as a mentor or trainer needs to be able to determine what kind of learner they are addressing and teach according to those needs.

The need for training is not limited to craftworkers. There must be supervisory training to ensure there are qualified leaders in the industry to supervise the craftworkers.

1.1.1 Motivation

As a supervisor or crew leader, it is important to understand what motivates your crew. Money is often thought to be a good motivator. Although that may be true to some extent, it has been proven to be a temporary solution. Once a person has reached a level of financial security, other factors come into play. Studies show that many people tend to be motivated by environment and conditions. For those people, a great workplace may provide more satisfaction than pay. If you give someone a raise, they tend to work harder for a period of time. Then the satisfaction dissipates and they may want another raise. People are often motivated by feeling a sense of accomplishment. That is why setting and working toward recognizable goals tends to make employees more productive. A person with a feeling of involvement or a sense of achievement is likely to be better motivated and help to motivate others.

1.1.2 Understanding Workers

Many older workers grew up in an environment in which they were taught to work hard and stay with the job until retirement. They expected to stay with a company for a long time, and companies were structured to create a family-type environment.

Times have changed. Younger workers have grown up in a highly mobile society and are used to rapid rewards. This generation of workers can sometimes be perceived as lazy and unmotivated, but in reality, they simply have a different perspective. For such workers, it may be better to give them small projects or break up large projects into smaller pieces so that they feel repetitively rewarded, thus enhancing their perception of success.

- *Goal setting* – Set short-term and long-term goals, including tasks to be done and expected time frames. Help the trainees understand that things can happen to offset the short-term goals. This is one reason to set long-term goals as well. Don't set them up for failure, as this leads to frustration, and frustration can lead to reduced productivity.
- *Feedback* – Timely feedback is important. For example, telling someone they did a good job last year, or criticizing them for a job they did a month ago, is meaningless. Simple recognition isn't always enough. Some type of reward should accompany positive feedback, even if it is simply recognizing the employee in a public way. Constructive feedback should be given in private and be accompanied by some positive action, such as providing one-on-one training to correct a problem.

1.1.3 Craft Training

Craft training is often informal, taking place on the job site, outside of a traditional training classroom. According to the American Society for Training and Development (ASTD), craft training is generally handled through on-the-job instruction by a qualified co-worker or conducted by a supervisor.

The Society of Human Resources Management (SHRM) offers the following tips to supervisors in charge of training their employees:

- *Help crew members establish career goals.* Once the goals are established, the training required to meet the goals can be readily identified.
- *Determine what kind of training to give.* Training can be on the job under the supervision of a co-worker. It can be one-on-one with the supervisor. It can involve cross-training to teach a new trade or skill, or it can involve delegating new or additional responsibilities.
- *Determine the trainee's preferred method of learning.* Some people learn best by watching, others from verbal instructions, and others by doing. When training more than one person at a time, try to use all three methods.

Communication is a critical component of training employees. The SHRM advises that supervisors do the following when training their employees:

- *Explain the task, why it needs to be done, and how it should be done.* Confirm that the trainees understand these three areas by asking questions. Allow them to ask questions as well.
- *Demonstrate the task.* Break the task down into manageable parts and cover one part at a time.
- *Ask trainees to do the task while you observe them.* Try not to interrupt them while they are doing the task unless they are doing something that is unsafe and potentially harmful.
- *Give the trainees feedback.* Be specific about what they did and mention any areas where they need to improve.

1.1.4 Supervisory Training

Given the need for skilled craftworkers and qualified supervisory personnel, it seems logical that companies would offer training to their employees through in-house classes, or by subsidizing outside training programs. While some contractors have their own in-house training programs or participate in training offered by associations and other organizations, many contractors do not offer training at all.

There are a number of reasons that companies do not develop or provide training programs, including the following:

- Lack of money to train
- Lack of time to train
- Lack of knowledge about the benefits of training programs
- High rate of employee turnover
- Workforce too small

- Past training involvement was ineffective
- The company hires only trained workers
- Lack of interest from workers
- Lack of company interest in training

For craftworkers to move up into supervisory and managerial positions, it will be necessary for them to continue their education and training. Those who are willing to acquire and develop new skills have the best chance of finding stable employment. It is therefore critical that employees take advantage of training opportunities, and that companies employ training as part of their business culture.

Your company has recognized the need for training. Your participation in a leadership training program such as this will begin to fill the gap between craft and supervisory training.

1.2.0 Impact of Technology

Many industries, including the construction industry, have made the move to technology as a means of remaining competitive. Benefits include increased productivity and speed, improved quality of documents, greater access to common data, and better financial controls and communication. As technology becomes a greater part of supervision, crew leaders need to be able to use it properly. One important concern with electronic communication is to keep it brief, factual, and legal. Because the receiver has no visual or auditory clues as to the sender's intent, the sender can be easily misunderstood. In other words, it is more difficult to tell if someone is just joking via e-mail because you can't see their face or hear the tone of their voice.

Cellular telephones, voicemail, and handheld communication devices have made it easy to keep in touch. They are particularly useful communication sources for contractors or crew leaders who are on a job site, away from their offices, or constantly on the go.

Cellular telephones allow the users to receive incoming calls as well as make outgoing calls. Unless the owner is out of the cellular provider's service area, the cell phone may be used any time to answer calls, make calls, and send and receive voicemail or email. Always check the company's policy with regard to the use of cell phones on the job.

Handheld communication devices known as smart phones allow supervisors to plan their calendars, schedule meetings, manage projects, and access their email from remote locations. These computers are small enough to fit in the palm of the hand, yet powerful enough to hold years of information from various projects. Information can be transmitted electronically to others on the project team or transferred to a computer.

2.0.0 GENDER AND CULTURAL ISSUES

During the past several years, the construction industry in the United States has experienced a shift in worker expectations and diversity. These two issues are converging at a rapid pace. At no time has there been such a generational merge in the workforce, ranging from The Silent Generation (1925–1945), Baby Boomers (1946–1964), Gen X (1965–1979), and the Millennials, also known as Generation Y (1980–2000).

This trend, combined with industry diversity initiatives, has created a climate in which companies recognize the need to embrace a diverse workforce that crosses generational, gender, and ethnic boundaries. To do this effectively, they are using their own resources, as well as relying on associations with the government and trade organizations. All current research indicates that industry will be more dependent on the critical skills of a diverse workforce—a workforce that is both culturally and ethnically fused. Across the United States, construction and other industries are aggressively seeking to bring new workers into their ranks, including women and racial and ethnic minorities. Diversity is no longer solely driven by social and political issues, but by consumers who need hospitals, malls, bridges, power plants, refineries, and many other commercial and residential structures.

Some issues relating to a diverse workforce will need to be addressed on the job site. These issues include different communication styles of men and women, language barriers associated with cultural differences, sexual harassment, and gender or racial discrimination.

2.1.0 Communication Styles of Men and Women

As more and more women move into construction, it becomes increasingly important that communication barriers between men and women are broken down and that differences in behaviors are understood so that men and women can work together more effectively. The Jamestown, New York Area Labor Management Committee (JALMC) offers the following explanations and tips:

- *Women tend to ask more questions than men do.* Men are more likely to proceed with a job and figure it out as they go along, while women are more likely to ask questions first.
- *Men tend to offer solutions before empathy; women tend to do the opposite.* Both men and women should say what they want up front, whether it's the solution to a problem, or simply a sympathetic ear. That way, both genders will feel understood and supported.
- *Women are more likely to ask for help when they need it.* Women are generally more pragmatic when it comes to completing a task. If they need help, they will ask for it. Men are more likely to attempt to complete a task by themselves, even when assistance is needed.
- *Men tend to communicate more competitively, and women tend to communicate more cooperatively.* Both parties need to hear one another out without interruption.

This does not mean that one method is more effective than the other. It simply means that men and women use different approaches to achieve the same result.

2.2.0 Language Barriers

Language barriers are a real workplace challenge for crew leaders. Millions of workers speak languages other than English. Spanish is commonly spoken in the United States. As the makeup of the immigrant population continues to change, the number of non-English speakers will rise dramatically, and the languages being spoken will also change. Bilingual job sites are increasingly common.

Companies have the following options to overcome this challenge:

- Offer English classes either at the work site or through school districts and community colleges.
- Offer incentives for workers to learn English.

As the workforce becomes more diverse, communicating with people for whom English is a second language will be even more critical. The following tips will help when communicating across language barriers:

- Be patient. Give workers time to process the information in a way that they can comprehend.
- Avoid humor. Humor is easily misunderstood and may be misinterpreted as a joke at the worker's expense.
- Don't assume that people are unintelligent simply because they don't understand what you are saying.
- Speak slowly and clearly, and avoid the tendency to raise your voice.
- Use face-to-face communication whenever possible. Over-the-phone communication is often more difficult when a language barrier is involved.
- Use pictures or drawings to get your point across.
- If a worker speaks English poorly but understands reasonably well, ask the worker to demonstrate his or her understanding through other means.

2.3.0 Cultural Differences

As workers from a multitude of backgrounds and cultures are brought together, there are bound to be differences and conflicts in the workplace.

To overcome cultural conflicts, the SHRM suggests the following approach to resolving cultural conflicts between individuals:

- *Define the problem from both points of view.* How does each person involved view the conflict? What does each person think is wrong? This involves moving beyond traditional thought processes to consider alternate ways of thinking.
- *Uncover cultural interpretations.* What assumptions are being made based on cultural programming? By doing this, the supervisor may realize what motivated an employee to act in a particular manner.

- *Create cultural synergy.* Devise a solution that works for both parties involved. The purpose is to recognize and respect other's cultural values, and work out mutually acceptable alternatives.

2.4.0 Sexual Harassment

In today's business world, men and women are working side-by-side in careers of all kinds. As women make the transition into traditionally male industries, such as construction, the likelihood of sexual harassment increases. Sexual harassment is defined as unwelcome behavior of a sexual nature that makes someone feel uncomfortable in the workplace by focusing attention on their gender instead of on their professional qualifications. Sexual harassment can range from telling an offensive joke or hanging a poster of a swimsuit-clad man or woman, to making sexual comments or physical advances.

Historically, sexual harassment was thought to be an act performed by men of power within an organization against women in subordinate positions. However, the number of sexual harassment cases over the years, have shown that this is no longer the case.

Sexual harassment can occur in a variety of circumstances, including but not limited to the following:

- The victim as well as the harasser may be a woman or a man. The victim does not have to be of the opposite sex.
- The harasser can be the victim's supervisor, an agent of the employer, a supervisor in another area, a co-worker, or a non-employee.
- The victim does not have to be the person harassed, but could be anyone affected by the offensive conduct.
- Unlawful sexual harassment may occur without economic injury to or discharge of the victim.
- The harasser's conduct must be unwelcome.

The Equal Employment Opportunity Commission (EEOC) enforces sexual harassment laws within industries. When investigating allegations of sexual harassment, the EEOC looks at the whole record, including the circumstances and the context in which the alleged incidents occurred. A decision on the allegations is made from the facts on a case-by-case basis. A crew leader who is aware of sexual harassment and does nothing to stop it can be held responsible. The crew leader therefore should not only take action to stop sexual harassment, but should serve as a good example for the rest of the crew.

Did you know?

Some companies employ what is known as sensitivity training in cases where individuals or groups have trouble adapting to a multi-cultural, multi-gender workforce. Sensitivity training is a psychological technique using group discussion, role playing, and other methods to allow participants to develop an awareness of themselves and how they interact with others.

Prevention is the best tool to eliminate sexual harassment in the workplace. The EEOC encourages employers to take steps to prevent sexual harassment from occurring. Employers should clearly communicate to employees that sexual harassment will not be tolerated. They do so by developing a policy on sexual harassment, establishing an effective complaint or grievance process, and taking immediate and appropriate action when an employee complains.

Both swearing and off-color remarks and jokes are not only offensive to co-workers, but also tarnish a worker's image. Crew leaders need to emphasize that abrasive or crude behavior may affect opportunities for advancement. If disciplinary action becomes necessary, it should be covered by company policy. A typical approach is a three-step process in which the perpetrator is first given a verbal reprimand. In the event of further violations, a written reprimand and warning are given. Dismissal typically accompanies subsequent violations.

2.5.0 Gender and Minority Discrimination

More attention is being placed on fair recruitment, equal pay for equal work, and promotions for women and minorities in the workplace. Consequently, many business practices, including the way employees are treated, the organization's hiring and promotional practices, and the way people are compensated, are being analyzed for equity.

Once a male-dominated industry, construction companies are moving away from this image and are actively recruiting and training women, younger workers, people from other cultures, and workers with disabilities. This means that organizations hire the best person for the job, without regard for race, sex, religion, age, etc.

To prevent discrimination cases, employers must have valid job-related criteria for hiring, compensation, and promotion. These measures must be used consistently for every applicant interview, employee performance appraisal, and hiring or promotion decision. Therefore, all workers responsible for recruitment, selection, and supervision of employees, and evaluating job performance, must be trained on how to use the job-related criteria legally and effectively.

3.0.0 BUSINESS ORGANIZATIONS

An organization is the relationship among the people within the company or project. The crew leader needs to be aware of two types of organizations. These are formal organizations and informal organizations.

A formal organization exists when the activities of the people within the work group are directed toward achieving a common goal. An example of a formal organization is a work crew consisting of four carpenters and two laborers led by a crew leader, all working together toward a common goal.

A formal organization is typically documented on an organizational chart, which outlines all the positions that make up an organization and shows how those positions are related. Some organizational charts even depict the people within each position and the person to whom they report, as well as the people that the person supervises. *Figures 1* and *2* show examples of organization charts for fictitious companies. Note that each of these positions represents an opportunity for advancement in the construction industry that a crew leader can eventually achieve.

An informal organization allows for communication among its members so they can perform as a group. It also establishes patterns of behavior that help them to work as a group, such as agreeing to use a specific training program.

An example of an informal organization is a trade association such as Associated Builders and Contractors (ABC), Associated General Contractors (AGC), and the National Association of Women in Construction (NAWIC). Those, along with the thousands of other trade associations in the U.S., provide a forum in which members with common concerns can share information, work on issues, and develop standards for their industry.

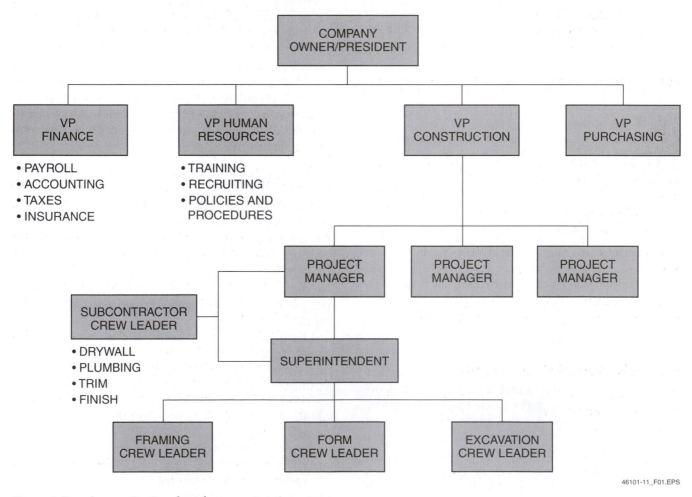

46101-11_F01.EPS

Figure 1 Sample organization chart for a construction company.

NCCER – *Contren® Learning Series*   46101-11

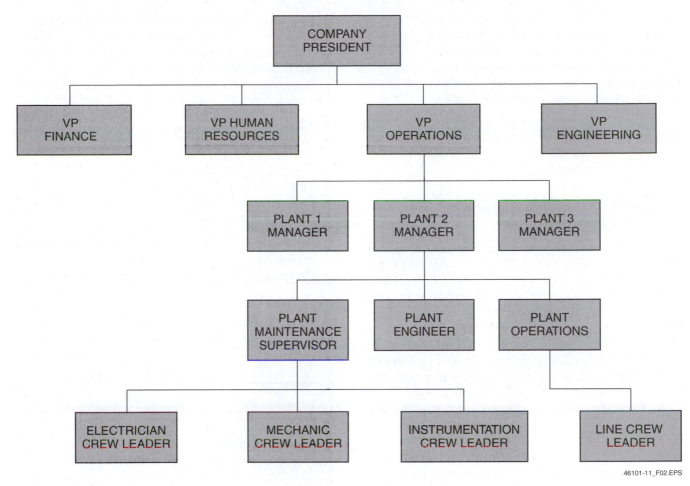

Figure 2 Sample organization chart for an industrial company.

Both types of organizations establish the foundation for how communication flows. The formal structure is the means used to delegate authority and responsibility and to exchange information. The informal structure is used to exchange information.

Members in an organization perform best when each member:

- Knows the job and how it will be done
- Communicates effectively with others in the group
- Understands his or her role in the group
- Recognizes who has the authority and responsibility

3.1.0 Division of Responsibility

The conduct of a business involves certain functions. In a small organization, responsibilities may be divided between one or two people. However, in a larger organization with many different and complex activities, responsibilities may be grouped into similar activity groups, and the responsibility for each group assigned to department managers. In either case, the following major departments exist in most companies:

- *Executive* – This office represents top management. It is responsible for the success of the company through short-range and long-range planning.
- *Human Resources* – This office is responsible for recruiting and screening prospective employees; managing employee benefits programs; advising management on pay and benefits; and developing and enforcing procedures related to hiring practices.
- *Accounting* – This office is responsible for all record keeping and financial transactions, including payroll, taxes, insurance, and audits.
- *Contract Administration* – This office prepares and executes contractual documents with owners, subcontractors, and suppliers.
- *Purchasing* – This office obtains material prices and then issues purchase orders. The purchasing office also obtains rental and leasing rates on equipment and tools.

- *Estimating*: This office is responsible for recording the quantity of material on the jobs, the takeoff, pricing labor and material, analyzing subcontractor bids, and bidding on projects.
- *Operations*: This office plans, controls, and supervises all project-related activities.

Other divisions of responsibility a company may create involve architectural and engineering design functions. These divisions usually become separate departments.

3.2.0 Authority, Responsibility, and Accountability

As an organization grows, the manager must ask others to perform many duties so that the manager can concentrate on management tasks. Managers typically assign (delegate) activities to their subordinates. When delegating activities, the crew leader assigns others the responsibility to perform the designated tasks.

Responsibility means obligation, so once the responsibility is delegated, the person to whom it is assigned is obligated to perform the duties.

Along with responsibility comes authority. *Authority* is the power to act or make decisions in carrying out an assignment. The type and amount of authority a supervisor or worker has depends on the company for which he or she works. Authority and responsibility must be balanced so employees can carry out their tasks. In addition, delegation of sufficient authority is needed to make an employee accountable to the crew leader for the results.

Accountability is the act of holding an employee responsible for completing the assigned activities. Even though authority and responsibility may be delegated to crew members, the final responsibility always rests with the crew leader.

3.3.0 Job Descriptions

Many companies furnish each employee with a written job description that explains the job in detail. Job descriptions set a standard for the employee. They make judging performance easier, clarify the tasks each person should handle, and simplify the training of new employees.

Each new employee should understand all the duties and responsibilities of the job after reviewing the job description. Thus, the time it takes for the employee to make the transition from being a new and uninformed employee to a more experienced member of a crew is shortened.

A job description need not be long, but it should be detailed enough to ensure there is no misunderstanding of the duties and responsibilities of the position. The job description should contain all the information necessary to evaluate the employee's performance.

A job description should contain, at minimum, the following:

- Job title
- General description of the position
- Minimum qualifications for the job
- Specific duties and responsibilities
- The supervisor to whom the position reports
- Other requirements, such as qualifications, certifications, and licenses

A sample job description is shown in *Figure 3*.

3.4.0 Policies and Procedures

Most companies have formal policies and procedures established to help the crew leader carry out his or her duties. A *policy* is a general state-

Position:
Crew Leader

General Summary:
First line of supervision on a construction crew installing concrete formwork.

Reports To:
Job Superintendent

Physical and Mental Responsibilities:
- Ability to stand for long periods
- Ability to solve basic math and geometry problems

Duties and Responsibilities:
- Oversee crew
- Provide instruction and training in construction tasks as needed
- Make sure proper materials and tools are on the site to accomplish tasks
- Keep project on schedule
- Enforce safety policies and procedures

Knowledge, Skills, and Experience Required:
- Extensive travel throughout the Eastern United States, home base in Atlanta
- Ability to operate a backhoe and trencher
- Valid commercial driver's license with no DUI violations
- Ability to work under deadlines with the knowledge and ability to foresee problem areas and develop a plan of action to solve the situation

46101-11_F03.EPS

Figure 3 Example of a job description.

ment establishing guidelines for a specific activity. Examples include policies on vacations, breaks, workplace safety, and checking out tools.

Procedures are the ways that policies are carried out. For example, a procedure written to implement a policy on workplace safety would include guidelines for reporting accidents and general safety procedures that all employees are expected to follow.

A crew leader must be familiar with the company policies and procedures, especially with regard to safety practices. When OSHA inspectors visit a job site, they often question employees and crew leaders about the company policies related to safety. If they are investigating an accident, they will want to verify that the responsible crew leader knew the applicable company policy and followed it.

Review Questions

1. The construction industry should provide training for craftworkers and supervisors _____.
   a. to ensure that there are enough future workers
   b. to avoid discrimination lawsuits
   c. in order to update the skills of older workers who are retiring at a later age than they previously did
   d. even though younger workers are now less likely to seek jobs in other areas than they were 10 years ago

2. Companies traditionally offer craftworker training _____.
   a. that a supervisor leads in a classroom setting
   b. that a craftworker leads in a classroom setting
   c. in a hands-on setting, where craftworkers learn from a co-worker or supervisor
   d. on a self-study basis to allow craftworkers to proceed at their own pace

3. One way to provide effective training is to _____.
   a. avoid giving negative feedback until trainees are more experienced in doing the task
   b. tailor the training to the career goals and needs of trainees
   c. choose one training method and use it for all trainees
   d. encourage trainees to listen, saving their questions for the end of the session

4. One way to prevent sexual harassment in the workplace is to _____.
   a. require employee training in which the potentially offensive subject of stereotypes is carefully avoided
   b. develop a consistent policy with appropriate consequences for engaging in sexual harassment
   c. communicate to workers that the victim of sexual harassment is the one who is being directly harassed, not those affected in a more indirect way
   d. educate workers to recognize sexual harassment for what it is—unwelcome conduct by the opposite sex

5. Employers can minimize all types of workplace discrimination by hiring based on a consistent list of job-related requirements.

    a. True
    b. False

6. Members tend to function best within an organization when they _____.

    a. are allowed to select their own style of clothing for each project
    b. understand their role
    c. do not disagree with the statements of other workers or supervisors
    d. are able to work without supervision

7. A formal organization is defined as a group of individuals who work independently, but share the same goal

    a. True
    b. False

8. A formal organization uses an organizational chart to _____.

    a. depict all companies with which it conducts business
    b. show all customers with which it conducts business
    c. track projects between departments
    d. show the relationships among the existing positions in the company

9. Which of the following is a function typically performed by the operations department of a company?

    a. Purchase materials
    b. Plan projects
    c. Prepare payrolls
    d. Recruiting and screening new hires

10. The company department that manages employee benefits and personnel recruiting is _____.

    a. Engineering
    b. Human Resources
    c. Purchasing
    d. Contract Administration

11. The power to make decisions and act on them in carrying out an assignment is _____.

    a. delegating
    b. responsibility
    c. decisiveness
    d. authority

12. Accountability is defined as _____.

    a. the power to act or make decisions in carrying out assignments
    b. giving an employee a particular task to perform
    c. the act of an employee responsible for the completion and results of a particular duty
    d. having the power to promote someone

13. A good job description should include _____.

    a. a complete organization chart
    b. any information needed to judge job performance
    c. the company dress code
    d. the company's sexual harassment policy

14. The purpose of a policy is to _____.

    a. establish company guidelines regarding a particular activity
    b. specify what tools and equipment are required for a job
    c. list all information necessary to judge an employee's performance
    d. inform employees about the future plans of the company

15. One example of a procedure would be the rules for taking time off.

    a. True
    b. False

Objectives

Upon completion of this section, you should be able to:

1. Describe the role of a crew leader.
2. List the characteristics of effective leaders.
3. Be able to discuss the importance of ethics in a supervisor's role.
4. Identify the three styles of leadership.
5. Describe the forms of communication.
6. Describe the four parts of verbal communication.
7. Describe the importance of active listening.
8. Explain how to overcome the barriers to communication.
9. List ways that leaders can motivate their employees.
10. Explain the importance of delegating and implementing policies and procedures.
11. Distinguish between problem solving and decision making.

1.0.0 INTRODUCTION TO LEADERSHIP

For the purpose of this program, it is important to define some of the positions that will be discussed. The term *craftworker* refers to a person who performs the work of his or her trade(s). The crew leader is a person who supervises one or more craftworkers on a crew. A superintendent is essentially an on-site supervisor who is responsible for one or more crew leaders or front-line supervisors. Finally, a project manager or general superintendent may be responsible for managing one or more projects. This training will concentrate primarily on the role of the crew leader.

Craftworkers and crew leaders differ in that the crew leader manages the activities that the skilled craftworkers on the crews actually perform. In order to manage a crew of craftworkers, a crew leader must have first-hand knowledge and experience in the activities being performed. In addition, he or she must be able to act directly in organizing and directing the activities of the various crew members.

This section explains the importance of developing effective leadership skills as a new crew leader. Effective ways to communicate with all levels of employees and co-workers, build teams, motivate crew members, make decisions, and resolve problems are covered in depth.

2.0.0 THE SHIFT IN WORK ACTIVITIES

The crew leader is generally selected and promoted from a work crew. The selection will often be based on that person's ability to accomplish tasks, to get along with others, to meet schedules, and to stay within the budget. The crew leader must lead the team to work safely and provide a quality product.

Making the transition from a craftworker to a crew leader can be difficult, especially when the new crew leader is in charge of supervising a group of peers. Crew leaders are no longer responsible for their work alone; rather, they are accountable for the work of an entire crew of people with varying skill levels and abilities, a multitude of personalities and work styles, and different cultural and educational backgrounds. Crew leaders must learn to put their personal relationships aside and work for the common goals of the team.

New crew leaders are often placed in charge of workers who were formerly their friends and peers on a crew. This situation can create some conflicts. For example, some of the crew may try to take advantage of the friendship by seeking special favors. They may also want to be privy to information that should be held closely. These problems can be overcome by working with the crew to set mutual performance goals and by freely communicating with them within permitted limits. Use their knowledge and strengths along with your own so that they feel like they are key players on the team.

As an employee moves from a craftworker position to the role of a crew leader, he or she will find that more hours will be spent supervising the work of others than actually performing the technical skill for which he or she has been trained. *Figure 4* represents the percentage of time craftworkers, crew leaders, superintendents, and project managers spend on technical and supervisory work as their management responsibilities increase.

The success of the new crew leader is directly related to the ability to make the transition from crew member into a leadership role.

3.0.0 BECOMING A LEADER

A crew leader must have leadership skills to be successful. Therefore, one of the primary goals of a person who wants to become a crew leader should be to develop strong leadership skills and learn to use them effectively.

There are many ways to define a leader. One straightforward definition is a person who influences other people in the achievement of a goal.

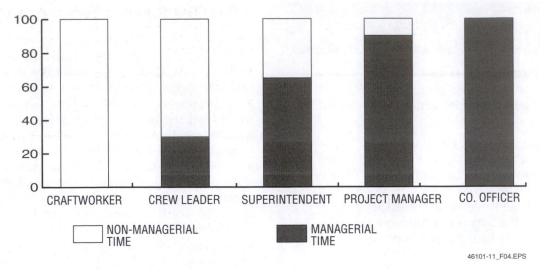

Figure 4 Percentage of time spent on technical and supervisory work.

Some people may have inherited leadership qualities or may have developed traits that motivate others to follow and perform. Research shows that people who possess such talents are likely to succeed as leaders.

3.1.0 Characteristics of Leaders

Leadership traits are similar to the skills that a crew leader needs in order to be effective. Although the characteristics of leadership are many, there are some definite commonalities among effective leaders.

First and foremost, effective leaders lead by example. In other words, they work and live by the standards that they establish for their crew members or followers, making sure they set a positive example.

Effective leaders also tend to have a high level of drive and determination, as well as a stick-to-it attitude. When faced with obstacles, effective leaders don't get discouraged; instead, they identify the potential problems, make plans to overcome them, and work toward achieving the intended goal. In the event of failure, effective leaders learn from their mistakes and apply that knowledge to future situations. They also learn from their successes.

Effective leaders are typically effective communicators who clearly express the goals of a project to their crew members. Accomplishing this may require that the leader overcome issues such as language barriers, gender bias, or differences in personalities to ensure that each member of the crew understands the established goals of the project.

Effective leaders have the ability to motivate their crew members to work to their full potential and become effective members of the team. Crew leaders try to develop crew member skills and encourage them to improve and learn as a means to contribute more to the team effort. Effective leaders strive for excellence from themselves and their team, so they work hard to provide the skills and leadership necessary to do so.

In addition, effective leaders must possess organizational skills. They know what needs to be accomplished, and they use their resources to make it happen. Because they cannot do it alone, leaders enlist the help of their team members to share in the workload. Effective leaders delegate work to their crew members, and they implement company policies and procedures to ensure that the work is completed safely, effectively, and efficiently.

Finally, effective leaders have the authority and self-confidence that allows them to make decisions and solve problems. In order to accomplish their goals, leaders must be able to calculate risks, absorb and interpret information, assess courses of action, make decisions, and assume the responsibility for those decisions.

3.1.1 Leadership Traits

There are many other traits of effective leaders. Some other major characteristics of leaders include the following:

- Ability to plan and organize
- Loyalty to their company and crew
- Ability to motivate
- Fairness

- Enthusiasm
- Willingness to learn from others
- Ability to teach others
- Initiative
- Ability to advocate an idea
- Good communication skills

3.1.2 Expected Leadership Behavior

Followers have expectations of their leaders. They look to their leaders to:

- Lead by example
- Suggest and direct
- Plan and organize the work
- Communicate effectively
- Make decisions and assume responsibility
- Have the necessary technical knowledge
- Be a loyal member of the group
- Abide by company policies and procedures

3.2.0 Functions of a Leader

The functions of a leader will vary with the environment, the group being led, and the tasks to be performed. However, there are certain functions common to all situations that the leader will be called upon to perform. Some of the major functions are:

- Organize, plan, staff, direct, and control work
- Empower group members to make decisions and take responsibility for their work
- Maintain a cohesive group by resolving tensions and differences among its members and between the group and those outside the group
- Ensure that all group members understand and abide by company policies and procedures
- Accept responsibility for the successes and failures of the group's performance
- Represent the group
- Be sensitive to the differences of a diverse workforce

3.3.0 Leadership Styles

There are three main styles of leadership. At one extreme is the autocratic or commander style of leadership, where the crew leader makes all of the decisions independently, without seeking the opinions of crew members. At the other extreme is the hands-off or facilitator style, where the crew leader empowers the employees to make decisions. In the middle is the democratic or collaborative style, where the crew leader seeks crew member opinions and makes the appropriate decisions based on their input.

The following are some characteristics of each of the three leadership styles:

Commander types:

- Expect crew members to work without questioning procedures
- Seldom seek advice from crew members
- Insist on solving problems alone
- Seldom permit crew members to assist each other
- Praise and criticize on a personal basis
- Have no sincere interest in creatively improving methods of operation or production

Partner types:

- Discuss problems with their crew members
- Listen to suggestions from crew members
- Explain and instruct
- Give crew members a feeling of accomplishment by commending them when they do a job well
- Are friendly and available to discuss personal and job-related problems

Facilitator types:

- Believe no supervision is best
- Rarely give orders
- Worry about whether they are liked by their crew members

Effective leadership takes many forms. The correct style for a particular situation or operation depends on the nature of the crew as well as the work it has to accomplish. For example, if the crew does not have enough experience for the job ahead, then a commander style may be appropriate. The autocratic style of leadership is also effective when jobs involve repetitive operations that require little decision-making.

However, if a worker's attitude is an issue, a partner style may be appropriate. In this case, providing the missing motivational factors may increase performance and result in the improvement of the worker's attitude. The democratic style of leadership is also used when the work is of a creative nature, because brainstorming and exchanging ideas with such crew members can be beneficial.

The facilitator style is effective with an experienced crew on a well-defined project.

The company must give a crew leader sufficient authority to do the job. This authority must be commensurate with responsibility, and it must be made known to crew members when they are hired so that they understand who is in charge.

A crew leader must have an expert knowledge of the activities to be supervised in order to be effective. This is important because the crew members need to know that they have someone to turn to when they have a question or a problem, when they need some guidance, or when modifications or changes are warranted by the job.

Respect is probably the most useful element of authority. Respect usually derives from being fair to employees, by listening to their complaints and suggestions, and by using incentives and rewards appropriately to motivate crew members. In addition, crew leaders who have a positive attitude and a favorable personality tend to gain the respect of their crew members as well as their peers. Along with respect comes a positive attitude from the crew members.

3.4.0 Ethics in Leadership

The crew leader should practice the highest standards of ethical conduct. Every day the crew leader has to make decisions that may have ethical implications. When an unethical decision is made, it not only hurts the crew leader, but also other workers, peers, and the company for which he or she works.

There are three basic types of ethics:

1. Business or legal
2. Professional or balanced
3. Situational

Business, or legal, ethics concerns adhering to all laws and regulations related to the issue.

Professional, or balanced, ethics relates to carrying out all activities in such a manner as to be honest and fair to everyone concerned.

Situational ethics pertains to specific activities or events that may initially appear to be a gray area. For example, you may ask yourself, "How will I feel about myself if my actions are published in the newspaper or if I have to justify my actions to my family, friends, and colleagues?"

The crew leader will often be put into a situation where he or she will need to assess the ethical consequences of an impending decision. For instance, should a crew leader continue to keep one of his or her crew working who has broken into a cold sweat due to overheated working conditions just because the superintendent says the activity is behind schedule? Or should a crew leader, who is the only one aware that the reinforcing steel placed by his or her crew was done incorrectly, correct the situation before the concrete is placed in the form? If a crew leader is ever asked to carry

through on an unethical decision, it is up to him or her to inform the superintendent of the unethical nature of the issue, and if still requested to follow through, refuse to act.

4.0.0 COMMUNICATION

Successful crew leaders learn to communicate effectively with people at all levels of the organization. In doing so, they develop an understanding of human behavior and acquire communication skills that enable them to understand and influence others.

There are many definitions of communication. Communication is the act of accurately and effectively conveying or transmitting facts, feelings, and opinions to another person. Simply stated, communication is the method of exchanging information and ideas.

Just as there are many definitions of communication, it also comes in many forms, including verbal, nonverbal, and written. Each of these forms of communication are discussed in this section.

4.1.0 Verbal Communication

Verbal communication refers to the spoken words exchanged between two or more people. Verbal communication consists of four distinct parts:

1. Sender
2. Message
3. Receiver
4. Feedback

Figure 5 depicts the relationship of these four parts within the communication process. In verbal communication, the focus is on feedback, which is used to verify that the sender's message was received as intended.

Did you know?

Research shows that the typical supervisor spends about 80 percent of his or her day communicating through writing, speaking, listening, or using body language. Of that time, studies suggest that approximately 20 percent of communication is written, and 80 percent involves speaking or listening.

Figure 5 Communication process.

4.1.1 The Sender

The sender is the person who creates the message to be communicated. In verbal communication, the sender actually says the message aloud to the person(s) for whom it is intended.

The sender must be sure to speak in a clear and concise manner that can be easily understood by others. This is not an easy task; it takes practice. Some basic speaking tips are:

- Avoid talking with anything in your mouth (food, gum, etc.).
- Avoid swearing and acronyms.
- Don't speak too quickly or too slowly. In extreme cases, people tend to focus on the rate of speech rather than what is being communicated.
- Pronounce words carefully to prevent misunderstandings.
- Speak with enthusiasm. Avoid speaking in a harsh voice or in a monotone.

4.1.2 The Message

The message is what the sender is attempting to communicate to the audience. A message can be a set of directions, an opinion, or a feeling. Whatever its function, a message is an idea or fact that the sender wants the audience to know.

Before speaking, determine what must be communicated, then take the time to organize what to say, ensuring that the message is logical and complete. Taking the time to clarify your thoughts prevents rambling, not getting the message across effectively, or confusing the audience. It also permits the sender to get to the point quickly.

In delivering the message, the sender should assess the audience. It is important not to talk down to them. Remember that everyone, whether in a senior or junior position, deserves respect and courtesy. Therefore, the sender should use words and phrases that the audience can understand and avoid technical language or slang. In addition, the sender should use short sentences, which gives the audience time to understand and digest one point or fact at a time.

4.1.3 The Receiver

The receiver is the person to whom the message is communicated. For the communication process to be successful, it is important that the receiver understands the message as the sender intended. Therefore, the receiver must listen to what is being said.

There are many barriers to effective listening, particularly on a busy construction job site. Some of these obstacles include the following:

- Noise, visitors, cell phones, or other distractions
- Preoccupation, being under pressure, or daydreaming
- Reacting emotionally to what is being communicated
- Thinking about how to respond instead of listening
- Giving an answer before the message is complete
- Personal biases to the sender's communication style
- Finishing the sender's sentence

Some tips for overcoming these barriers are:

- Take steps to minimize or remove distractions; learn to tune out your surroundings
- Listen for key points
- Take notes
- Try not to take things personally
- Allow yourself time to process your thoughts before responding
- Let the sender communicate the message without interruption
- Be aware of your personal biases, and try to stay open-minded

There are many ways for a receiver to show that he or she is actively listening to what is being said. This can even be accomplished without saying a word. Examples include maintaining eye contact, nodding your head, and taking notes. It may also be accomplished through feedback.

4.1.4 Feedback

Feedback refers to the communication that occurs after the message has been sent by the sender and received by the receiver. It involves the receiver responding to the message.

Feedback is a very important part of the communication process because it allows the receiver to communicate how he or she interpreted the message. It also allows the sender to ensure that the message was understood as intended. In other words, feedback is a checkpoint to make sure the receiver and sender are on the same page.

The receiver can use the opportunity of providing feedback to paraphrase back what was heard. When paraphrasing what you heard, it is best to use your own words. That way, you can show the sender that you interpreted the message correctly and could explain it to others if needed.

In addition, the receiver can clarify the meaning of the message and request additional information when providing feedback. This is generally accomplished by asking questions.

One opportunity to provide feedback is in the performance of crew evaluations. Many companies have formal evaluation forms that are used on a yearly basis to evaluate workers for pay increases. These evaluations should not come as a once-a-year surprise. An effective crew leader provides constant performance feedback, which is ultimately reflected in the annual performance evaluation. It is also important to stress the importance of self-evaluation with your crew.

4.2.0 Nonverbal Communication

Unlike verbal or written communication, nonverbal communication does not involve the spoken or written word. Rather, non-verbal communication refers to things that you can actually see when communicating with others. Examples include facial expressions, body movements, hand gestures, and eye contact.

Nonverbal communication can provide an external signal of an individual's inner emotions. It occurs simultaneously with verbal communication; often, the sender of the nonverbal communication is not even aware of it.

Because it can be physically observed, nonverbal communication is just as important as the words used in conveying the message. Often, people are influenced more by nonverbal signals than by spoken words. Therefore, it is important to be conscious of nonverbal cues because you don't want the receiver to interpret your message incorrectly based on your posture or an expression on your face. After all, these things may have nothing to do with the communication exchange; instead, they may be carrying over from something else going on in your day.

4.3.0 Written or Visual Communication

Some communication will have to be written or visual. Written or visual communication refers to communication that is documented on paper or transmitted electronically using words or visuals.

Many messages on a job have to be communicated in text form. Examples include weekly reports, requests for changes, purchase orders, and correspondence on a specific subject. These items are written because they must be recorded for business and historical purposes. In addition, some communication on the job will have to be visual. Items that are difficult to explain verbally or by the written word can best be explained through diagrams or graphics. Examples include the plans or drawings used on a job.

When writing or creating a visual message, it is best to assess the reader or the audience before beginning. The reader must be able to read the message and understand the content; otherwise, the communication process will be unsuccessful. Therefore, the writer should consider the actual meaning of words or diagrams and how others might interpret them. In addition, the writer should make sure that all handwriting is legible if the message is being handwritten.

Here are some basic tips for writing:

- Avoid emotion-packed words or phrases.
- Be positive whenever possible.
- Avoid using technical language or jargon.
- Stick to the facts.
- Provide an adequate level of detail.
- Present the information in a logical manner.
- Avoid making judgments unless asked to do so.
- Proofread your work; check for spelling and grammatical errors.
- Make sure that the document is legible.
- Avoid using acronyms.
- Make sure the purpose of the message is clearly stated.
- Be prepared to provide a verbal or visual explanation, if needed.

Here are some basic tips for creating visuals:

- Provide an adequate level of detail.
- Ensure that the diagram is large enough to be seen.
- Avoid creating complex visuals; simplicity is better.
- Present the information in a logical order.
- Be prepared to provide a written or verbal explanation of the visual, if needed.

4.4.0 Communication Issues

It is important to note that each person communicates a little differently; that is what makes us unique as individuals. As the diversity of the workforce changes, communication will become even more challenging because the audience may include individuals from different ethnic groups, cultural backgrounds, educational levels, and economic status groups. Therefore, it is necessary to assess the audience in order to determine how to communicate effectively with each individual.

The key to effective communication is to acknowledge that people are different and to be able to adjust the communication style to meet the needs of the audience or the person on the receiving end of your message. This involves relaying the message in the simplest way possible, avoiding the use of words that people may find confusing. Be aware of how you use technical language, slang, jargon, and words that have multiple meanings. Present the information in a clear, concise manner. Avoid rambling and always speak clearly, using good grammar.

In addition, be prepared to communicate the message in multiple ways or adjust your level of detail or terminology to ensure that everyone understands the meaning as intended. For instance, a visual person who cannot comprehend directions in a verbal or written form may need a map. It may be necessary to overcome language barriers on the job site by using graphics or visual aids to relay the message.

Figure 6 shows how to tailor the message to the audience.

VERBAL INSTRUCTIONS Experienced Crew	VERBAL INSTRUCTIONS Newer Crew	WRITTEN INSTRUCTIONS	DIAGRAM/MAP
"Please drive to the supply shop to pick up our order."	"Please drive to the supply shop. Turn right here and left at Route 1. It's at 75th Street and Route 1. Tell them the company name and that you're there to pick up our order."	1. Turn right at exit. 2. Drive 2 miles to Route 1. Turn LEFT. 3. Drive 1 mile (pass the tire shop) to 75th Street. 4. Look for supply store on right. . . .	

Different people learn in different ways. Be sure to communicate so you can be understood.

46101-11_F06.EPS

Figure 6 Tailor your message.

Read the following verbal conversations, and identify any problems:

Conversation I:

Judy: Hey, Roger…

Roger: What's up?

Judy: Has the site been prepared for the job trailer yet?

Roger: Job trailer?

Judy: The job trailer—it's coming in today. What time will the job site be prepared?

Roger: The trailer will be here about 1:00 PM.

Judy: The job site! What time will the job site be prepared?

_____

_____

_____

_____

_____

Conversation II:

John: Hey, Mike, I need your help.

Mike: What is it?

John: You and Joey go over and help Al's crew finish laying out the site.

Mike: Why me? I can't work with Joey. He can't understand a word I say.

John: Al's crew needs some help, and you and Joey are the most qualified to do the job.

Mike: I told you, I can't work with Joey.

_____

_____

_____

_____

_____

Conversation III:

Ed: Hey, Jill.

Jill: Sir?

Ed: Have you received the latest DOL, EEO requirement to be sure the OFCP administrator finds our records up to date when he reviews them in August?

Jill: DOL, EEO, and OFCP?

Ed: Oh, and don't forget the MSHA, OSHA, and EPA reports are due this afternoon.

Jill: MSHA, OSHA, and EPA?

_____

_____

_____

_____

_____

_____

_____

Conversation IV:

Susan: Hey, Bob, would you do me a favor?

Bob: Okay, Sue. What is it?

Susan: I was reading the concrete inspection report and found the concrete in Bays 4A, 3B, 6C, and 5D didn't meet the 3,000 psi strength requirements. Also, the concrete inspector on the job told me the two batches that came in today had to be refused because they didn't meet the slump requirements as noted on page 16 of the spec. I need to know if any placement problems happened on those bays, how long the ready mix trucks were waiting today, and what we plan to do to stop these problems in the future.

_____

_____

_____

_____

_____

Read the following written memos, and identify any problems:

Memo I:

Let's start with the transformer vault $285.00 due. For what you ask? Answer: practically nothing I admit, but here is the story. Paul the superintendent decided it was not the way good ole Comm Ed wanted it, we took out the ladder and part of the grading (as Paul instructed us to do) we brought it back here to change it. When Comm Ed the architect or Doe found out that everything would still work the way it was, Paul instructed us to reinstall the work. That is the whole story there is please add the $285.00 to my next payout.

_____

_____

_____

_____

_____

_____

Memo II:

Let's take rooms C 307-C-312 and C-313 we made the light track supports and took them to the job to erect them when we tried to put them in we found direct work in the way, my men spent all day trying to find out what to do so ask your Superintendent (Frank) he will verify seven hours pay for these men as he went back and forth while my men waited. Now the Architect has changed the system of hanging and has the gall to say that he has made my work easier, I can't see how. Anyway, we want an extra two (2) men for seven (7) hours for April 21 at $55.00 per hour or $385.00 on April 28th Doe Reference 197 finally resolved this problem. We will have no additional charges on Doe Reference 197, please note.

_____

_____

_____

_____

_____

5.0.0 MOTIVATION

The ability to motivate others is a key skill that leaders must develop. Motivation is the ability to influence. It also describes the amount of effort that a person is willing to put forth to accomplish something. For example, a crew member who skips breaks and lunch in an effort to complete a job on time is thought to be highly motivated, but a crew member who does the bare minimum or just enough to keep his or her job is considered unmotivated.

Employee motivation has dimension because it can be measured. Examples of how motivation can be measured include determining the level of absenteeism, the percentage of employee turnover, and the number of complaints, as well as the quality and quantity of work produced.

5.1.0 Employee Motivators

Different things motivate different people in different ways. Consequently, there is no one-size-fits-all approach to motivating crew members. It is im-portant to recognize that what motivates one crew member may not motivate another. In addition, what works to motivate a crew member once may not motivate that same person again in the future.

Frequently, the needs that motivate individuals are the same as those that create job satisfaction. They include the following:

- Recognition and praise
- Accomplishment
- Opportunity for advancement
- Job importance
- Change
- Personal growth
- Rewards

A crew leader's ability to satisfy these needs increases the likelihood of high morale within a crew. Morale refers to an individual's attitude toward the tasks he or she is expected to perform. High morale, in turn, means that employees will be motivated to work hard, and they will have a positive attitude about coming to work and doing their jobs.

5.1.1 Recognition and Praise

Recognition and praise refer to the need to have good work appreciated, applauded, and acknowledged by others. This can be accomplished by simply thanking employees for helping out on a project, or it can entail more formal praise, such as an award for Employee of the Month.

Some tips for giving recognition and praise include the following:

- Be available on the job site so that you have the opportunity to witness good work.
- Know good work and praise it when you see it.
- Look for good work and look for ways to praise it.
- Give recognition and praise only when truly deserved; otherwise, it will lose its meaning.
- Acknowledge satisfactory performance, and encourage improvement by showing confidence in the ability of the crew members to do above-average work.

5.1.2 Accomplishment

Accomplishment refers to a worker's need to set challenging goals and achieve them. There is nothing quite like the feeling of achieving a goal, particularly a goal one never expected to accomplish in the first place.

Crew leaders can help their crew members attain a sense of accomplishment by encouraging them to develop performance plans, such as goals for the year that will be used in performance evaluations. In addition, crew leaders can provide the support and tools (such as training and coaching) necessary to help their crew members achieve these goals.

5.1.3 Opportunity for Advancement

Opportunity for advancement refers to an employee's need to gain additional responsibility and develop new skills and abilities. It is important that employees know that they are not limited to their current jobs. Let them know that they have a chance to grow with the company and to be promoted as recognition for excelling in their work.

Effective leaders encourage their crew members to work to their full potentials. In addition, they share information and skills with their employees in an effort to help them to advance within the organization.

5.1.4 Job Importance

Job importance refers to an employee's need to feel that his or her skills and abilities are valued and make a difference. Employees who do not feel valued tend to have performance and attendance issues. Crew leaders should attempt to make every crew member feel like an important part of the team, as if the job wouldn't be possible without their help.

5.1.5 Change

Change refers to an employee's need to have variety in work assignments. Change is what keeps things interesting or challenging. It prevents the boredom that results from doing the same task day after day with no variety.

5.1.6 Personal Growth

Personal growth refers to an employee's need to learn new skills, enhance abilities, and grow as a person. It can be very rewarding to master a new competency on the job. Similar to change, personal growth prevents the boredom associated with doing the same thing day after day without developing any new skills.

Crew leaders should encourage the personal growth of their employees as well as themselves. Learning should be a two-way street on the job site; crew leaders should teach their crew members and learn from them as well. In addition, crew members should be encouraged to learn from each other.

5.1.7 Rewards

Rewards are compensation for hard work. Rewards can include a crew member's base salary or go beyond that to include bonuses or other incentives. They can be monetary in nature (salary raises, holiday bonuses, etc.), or they can be non-monetary, such as free merchandise (shirts, coffee mugs, jackets, etc.) or other prizes. Attendance at training courses can be another form of reward.

5.2.0 Motivating Employees

To increase motivation in the workplace, crew leaders must individualize how they motivate different crew members. It is important that crew leaders get to know their crew members and determine what motivates them as individuals. Once again, as diversity increases in the workforce, this becomes even more challenging; therefore, effective communication skills are essential.

Here is a list of some tips for motivating employees:

- Keep jobs challenging and interesting. Boredom is a guaranteed de-motivator.
- Communicate your expectations. People need clear goals in order to feel a sense of accomplishment when the goals are achieved.
- Involve the employees. Feeling that their opinions are valued leads to pride in ownership and active participation.
- Provide sufficient training. Give employees the skills and abilities they need to be motivated to perform.
- Mentor the employees. Coaching and supporting employees boosts their self-esteem, their self-confidence, and ultimately their motivation.
- Lead by example. Become the kind of leader employees admire and respect, and they will be motivated to work for you.
- Treat employees well. Be considerate, kind, caring, and respectful; treat employees the way that you want to be treated.
- Avoid using scare tactics. Threatening employees with negative consequences can backfire, resulting in employee turnover instead of motivation.
- Reward your crew for doing their best by giving them easier tasks from time to time. It is tempting to give your best employees the hardest or dirtiest jobs because you know they will do the jobs correctly.
- Reward employees for a job well done.

6.0.0 TEAM BUILDING

Organizations are making the shift from the traditional boss-worker mentality to one that promotes teamwork. The manager becomes the team leader, and the workers become team members. They all work together to achieve the common goals of the team.

There are a number of benefits associated with teamwork. They include the ability to complete complex projects more quickly and effectively, higher employee satisfaction, and a reduction in turnover.

6.1.0 Successful Teams

Successful teams are made up of individuals who are willing to share their time and talents in an effort to reach a common goal—the goal of the team. Members of successful teams possess an *Us* or *We* attitude rather than an *I* or *You* attitude; they consider what's best for the team and put their egos aside.

Some characteristics of successful teams include the following:

- Everyone participates and every team member counts.
- There is a sense of mutual trust and interdependence.
- Team members are empowered.
- They communicate.
- They are creative and willing to take risks.
- The team leader develops strong people skills and is committed to the team.

6.2.0 Building Successful Teams

To be successful in the team leadership role, the crew leader should contribute to a positive attitude within the team.

There are several ways in which the team leader can accomplish this. First, he or she can work with the team members to create a vision or purpose of what the team is to achieve. It is important that every team member is committed to the purpose of the team, and the team leader is instrumental in making this happen.

Team leaders within the construction industry are typically assigned a crew. However, it can be beneficial for the team leader to be involved in selecting the team members. Selection should be based on a willingness of people to work on the team and the resources that they are able to bring to the team.

When forming a new team, team leaders should do the following:

- Explain the purpose of the team. Team members need to know what they will be doing, how long they will be doing it (if they are temporary or permanent), and why they are needed.
- Help the team establish goals or targets. Teams need a purpose, and they need to know what it is they are responsible for accomplishing.
- Define team member roles and expectations. Team members need to know how they fit into the team and what is expected of them as members of the team.
- Plan to transfer responsibility to the team as appropriate. Teams should be responsible for the tasks to be accomplished.

7.0.0 GETTING THE JOB DONE

Crew leaders must implement policies and procedures to make sure that the work is done correctly. Construction jobs have crews of people with various experiences and skill levels available to perform the work. The crew leader's job is to draw from this expertise to get the job done well and in a timely manner.

7.1.0 Delegating

Once the various activities that make up the job have been determined, the crew leader must identify the person or persons who will be responsible for completing each activity. This requires that the crew leader be aware of the skills and abilities of the people on the crew. Then, the crew leader must put this knowledge to work in matching the crew's skills and abilities to specific tasks that must be accomplished to complete the job.

After matching crew members to specific activities, the crew leader must then delegate the assignments to the responsible person(s). Delegation is generally communicated verbally by the crew leader talking directly to the person who has been assigned the activity. However, there may be times when work is assigned indirectly through written instructions or verbally through someone other than the crew leader.

When delegating work, remember to:

- Delegate work to a crew member who can do the job properly. If it becomes evident that he or she does not perform to the standard desired, either teach the crew member to do the work correctly or turn it over to someone else who can.

- Make sure the crew member understands what to do and the level of responsibility. Make sure desired results are clear, specify the boundaries and deadlines for accomplishing the results, and note the available resources.
- Identify the standards and methods of measurement for progress and accomplishment, along with the consequences of not achieving the desired results. Discuss the task with the crew member and check for understanding by asking questions. Allow the crew member to contribute feedback or make suggestions about how the task should be performed in a safe and quality manner.
- Give the crew member the time and freedom to get started without feeling the pressure of too much supervision. When making the work assignment, be sure to tell the crew member how much time there is to complete it, and confirm that this time is consistent with the job schedule.
- Examine and evaluate the result once a task is complete. Then, give the crew member some feedback as to how well it has been done. Get the crew member's comments. The information obtained from this is valuable and will enable the crew leader to know what kind of work to assign that crew member in the future. It will also give the crew leader a means of measuring his or her own effectiveness in delegating work.

7.2.0 Implementing Policies and Procedures

Every company establishes policies and procedures that employees are expected to follow and the crew leaders are expected to implement. Company policies and procedures are essentially guides for how the organization does business. They can also reflect organizational philosophies such as putting safety first or making the customer the top priority. Examples of policies and procedures include safety guidelines, credit standards, and billing processes.

Here are some tips for implementing policies and procedures:

- Learn the purpose of each policy. That way, you can follow it and apply it properly and fairly.
- If you're not sure how to apply a company policy or procedure, check the company manual or ask your supervisor.
- Apply company policies and procedures. Remember that they combine what's best for the customer and the company. In addition, they provide direction on how to handle specific situations and answer questions.

- If you are uncertain how to apply a policy, check with your supervisor.

Crew leaders may need to issue orders to their crew members. Basically, an order initiates, changes, or stops an activity. Orders may be general or specific, written or oral, and formal or informal. The decision of how an order will be issued is up to the crew leader, but it is governed by the policies and procedures established by the company.

When issuing orders:

- Make them as specific as possible.
- Avoid being general or vague unless it is impossible to foresee all of the circumstances that could occur in carrying out the order.
- Recognize that it is not necessary to write orders for simple tasks unless the company requires that all orders be written.
- Write orders for more complex tasks that will take considerable time to complete or orders that are permanent.
- Consider what is being said, the audience to whom it applies, and the situation under which it will be implemented to determine the appropriate level of formality for the order.

8.0.0 PROBLEM SOLVING AND DECISION MAKING

Problem solving and decision making are a large part of every crew leader's daily work. There will always be problems to be resolved and decisions to be made, especially in fast-paced, deadline-oriented industries.

8.1.0 Decision Making vs. Problem Solving

Sometimes, the difference between decision making and problem solving is not clear. Decision making refers to the process of choosing an alternative course of action in a manner appropriate for the situation. Problem solving involves determining the difference between the way things are and the way things should be, and finding out how to bring the two together. The two activities are interrelated because in order to make a decision, you may also have to use problem-solving techniques.

8.2.0 Types of Decisions

Some decisions are routine or simple. Such decisions can be made based on past experiences. An example would be deciding how to get to and from work. If you've worked at the same place for a long time, you are already aware of the options for traveling to and from work (take the bus, drive a car, carpool with a co-worker, take a taxi, etc.). Based on past experiences with the options identified, you can make a decision about how best to get to and from work.

On the other hand, some decisions are more difficult. These decisions require more careful thought about how to carry out an activity by using a formal problem-solving technique. An example is planning a trip to a new vacation spot. If you are not sure how to get there, where to stay, what to see, etc., one option is to research the area to determine the possible routes, hotel accommodations, and attractions. Then, you will have to make a decision about which route to take, what hotel to choose, and what sites to visit, without the benefit of direct past experience.

8.3.0 Problem Solving

The ability to solve problems is an important skill in any workplace. It's especially important for craftworkers, whose workday is often not predictable or routine. In this section, you will learn a five-step process for solving problems, which you can apply to both workplace and personal issues. Review the following steps and then see how they can be applied to a job-related problem. Keep in mind that a problem will not be solved until everyone involved admits that there is a problem.

Step 1 **Define the problem.** This isn't as easy as it sounds. Thinking through the problem often uncovers additional problems.

Step 2 **Think about different ways to solve the problem.** There is often more than one solution to a problem, so you must think through each possible solution and pick the best one. The best solution might be taking parts of two different solutions and combining them to create a new solution.

Step 3 **Pick the solution that seems best and figure out an action plan.** It is best to receive input both from those most affected by the problem and from those who will be most affected by any potential solution.

Step 4 **Test the solution to determine whether it actually works.** Many solutions sound great in theory but in practice don't turn out to be effective. On the other hand, you might discover from trying to apply

a solution that it is acceptable with a little modification. If a solution does not work, think about how you could improve it, and then implement your new plan.

Step 5 **Evaluate the process.** Review the steps you took to discover and implement the solution. Could you have done anything better? If the solution turns out to be satisfactory, you can add the solution to your knowledge base.

Next, you will see how to apply the problem-solving process to a workplace problem. Read the following situation and apply the five-step problem-solving process to come up with a solution to the issues posed by the situation.

Situation:

You are part of a team of workers assigned to a new shopping mall project. The project will take about 18 months to complete. The only available parking is half a mile from the job site. The crew has to carry heavy toolboxes and safety equipment from their cars and trucks to the work area at the start of the day, and then carry them back at the end of their shifts.

Step 1 **Define the problem.** Workers are wasting time and energy hauling all their equipment to and from the work site.

Step 2 **Think about different ways to solve the problem.** Several solutions have been proposed:
- Install lockers for tools and equipment closer to the work site.
- Have workers drive up to the work site to drop off their tools and equipment before parking.
- Bring in another construction trailer where workers can store their tools and equipment for the duration of the project.
- Provide a round-trip shuttle service to ferry workers and their tools.

NOTE

Each solution will have pros and cons, so it is important to receive input from the workers affected by the problem. For example, workers will probably object to any plan (like the drop-off plan) that leaves their tools vulnerable to theft.

Step 3 ***Pick the solution that seems best and figure out an action plan.*** The workers decide that the shuttle service makes the most sense. It should solve the time and energy problem, and workers can keep their tools with them. To put the plan into effect, the project supervisor arranges for a large van and driver to provide the shuttle service.

Step 4 ***Test the solution to determine whether it actually works.*** The solution works, but there is a problem. All the workers are scheduled to start and leave at the same time, so there is not enough room in the van for all the workers and their equipment. To solve this problem, the supervisor schedules trips spaced 15 minutes apart. The supervisor also adjusts worker schedules to correspond with the trips. That way, all the workers will not try to get on the shuttle at the same time.

Step 6 ***Evaluate the process.*** This process gave both management and workers a chance to express an opinion and discuss the various solutions. Everyone feels pleased with the process and the solution.

8.4.0 Special Leadership Problems

Because they are responsible for leading others, it is inevitable that crew leaders will encounter problems and be forced to make decisions about how to respond to the problem. Some problems will be relatively simple to resolve, like covering for a sick crew member who has taken a day off from work. Other problems will be complex and much more difficult to handle.

Some complex problems are relatively common. A few of the major employee problems include:

- Inability to work with others
- Absenteeism and turnover
- Failure to comply with company policies and procedures

8.4.1 Inability to Work with Others

Crew leaders will sometimes encounter situations where an employee has a difficult time working with others on the crew. This could be a result of personality differences, an inability to communicate, or some other cause. Whatever the reason, the crew leader must address the issue and get the crew working as a team.

The best way to determine the reason for why individuals don't get along or work well together is to talk to the parties involved. The crew leader should speak openly with the employee, as well as the other individual(s) to uncover the source of the problem and discuss its resolution.

Once the reason for the conflict is found, the crew leader can determine how to respond. There may be a way to resolve the problem and get the workers communicating and working as a team again. On the other hand, there may be nothing that can be done that will lead to a harmonious solution. In this case, the crew leader would either have to transfer the employee to another crew or have the problem crew member terminated. This latter option should be used as a last measure and should be discussed with one's superiors or Human Resources Department.

8.4.2 Absenteeism and Turnover

Absenteeism and turnover are big problems. Without workers available to do the work, jobs are delayed, and money is lost.

Absenteeism refers to workers missing their scheduled work time on a job. Absenteeism has many causes, some of which are inevitable. For instance, people get sick, they have to take time off for family emergencies, and they have to attend family events such as funerals. However, there are some causes of absenteeism that can be prevented by the crew leader.

The most effective way to control absenteeism is to make the company's policy clear to all employees. Companies that do this find that chronic absenteeism is reduced. New employees should have the policy explained to them. This explanation should include the number of absences allowed and the reasons for which sick or personal days can be taken. In addition, all workers should know how to inform their crew leaders when they miss work and understand the consequences of exceeding the number of sick or personal days allowed.

Once the policy on absenteeism is explained to employees, crew leaders must be sure to implement it consistently and fairly. If the policy is administered equally, employees will likely follow it. However, if the policy is not administered equally and some employees are given exceptions, then it will not be effective. Consequently, the rate of absenteeism is likely to increase.

Despite having a policy on absenteeism, there will always be employees who are chronically late or miss work. In cases where an employee abuses the absenteeism policy, the crew leader should discuss the situation directly with the employee. The crew leader should confirm that the employee understands the company's policy and insist that the employee comply with it. If the employee's behavior continues, disciplinary action may be in order.

Turnover refers to the loss of an employee that is initiated by that employee. In other words, the employee quits and leaves the company to work elsewhere or is fired for cause.

Like absenteeism, there are some causes of turnover that cannot be prevented and others that can. For instance, it is unlikely that a crew leader could keep an employee who finds a job elsewhere earning twice as much money. However, crew leaders can prevent some employee turnover situations. They can work to ensure safe working conditions for their crew, treat their workers fairly and consistently, and help promote good working conditions. The key is communication. Crew leaders need to know the problems if they are going to be able to successfully resolve them.

Some of the major causes of turnover include the following:

- Unfair/inconsistent treatment by the immediate supervisor
- Unsafe project sites
- Lack of job security

For the most part, the actions described for absenteeism are also effective for reducing turnover. Past studies have shown that maintaining harmonious relationships on the job site goes a long way in reducing both turnover and absenteeism. This requires effective leadership on the part of the crew leader.

8.4.3 Failure to Comply With Company Policies and Procedures

Policies are the rules that define the relationship between the company, its employees, its clients, and its subcontractors. Procedures are the instructions for carrying out the policies. Some companies have dress codes that are reflected in their policies. The dress code may be partly to ensure safety, and partly to define the image a company wants to project to the outside world.

Companies develop procedures to ensure that everyone who performs a task does it safely and efficiently. Many procedures deal with safety. A lockout/tagout procedure is an example. In this procedure, the company defines who may perform a lockout, how it is done, and who has the authority to remove or override it. Workers who fail to follow the procedure endanger themselves, as well as their co-workers.

Among a typical company's policies is the policy on disciplinary action. This policy defines steps to be taken in the event that an employee violates the company's policies or procedures. The steps range from counseling by a supervisor for the first offense, to a written warning, to dismissal for repeat offenses. This will vary from one company to another. For example, some companies will fire an employee for any violation of safety procedures.

The crew leader has the first-line responsibility for enforcing company policies and procedures. The crew leader should take the time with a new crew member to discuss the policies and procedures and show the crew member how to access them. If a crew member shows a tendency to neglect a policy or procedure, it is up to the crew leader to counsel that individual. If the crew member continues to violate a policy or procedure, the crew leader has no choice but to refer that individual to the appropriate authority within the company for disciplinary action.

Case I:

On the way over to the job trailer, you look up and see a piece of falling scrap heading for one of the laborers. Before you can say anything, the scrap material hits the ground about five feet in front of the worker. You notice the scrap is a piece of conduit. You quickly pick it up, assuring the worker you will take care of this matter.

Looking up, you see your crew on the third floor in the area from which the material fell. You decide to have a talk with them. Once on the deck, you ask the crew if any of them dropped the scrap. The men look over at Bob, one of the electricians in your crew. Bob replies, "I guess it was mine. It slipped out of my hand."

It is a known fact that the Occupational Safety and Health Administration (OSHA) regulations state that an enclosed chute of wood shall be used for material waste transportation from heights of 20 feet or more. It is also known that Bob and the laborer who was almost hit have been seen arguing lately.

1. Assuming Bob's action was deliberate, what action would you take?

_____

_____

_____

2. Assuming the conduit accidentally slipped from Bob's hand, how can you motivate him to be more careful?

_____

_____

_____

3. What follow-up actions, if any, should be taken relative to the laborer who was almost hit?

_____

_____

_____

4. Should you discuss the apparent OSHA violation with the crew? Why or why not?

_____

_____

_____

5. What acts of leadership would be effective in this case? To what leadership traits are they related?

_____

_____

_____

Case II:

Mike has just been appointed crew leader of a tile-setting crew. Before his promotion into management, he had been a tile setter for five years. His work had been consistently of superior quality.

Except for a little good-natured kidding, Mike's co-workers had wished him well in his new job. During the first two weeks, most of them had been cooperative while Mike was adjusting to his supervisory role.

At the end of the second week, a disturbing incident took place. Having just completed some of his duties, Mike stopped by the job-site wash station. There he saw Steve and Ron, two of his old friends who were also in his crew, washing.

"Hey, Ron, Steve, you should not be cleaning up this soon. It's at least another thirty minutes until quitting time," said Mike. "Get back to your work station, and I'll forget I saw you here."

"Come off it, Mike," said Steve. "You used to slip up here early on Fridays. Just because you have a little rank now, don't think you can get tough with us." To this Mike replied, "Things are different now. Both of you get back to work, or I'll make trouble." Steve and Ron said nothing more, and they both returned to their work stations.

From that time on, Mike began to have trouble as a crew leader. Steve and Ron gave him the silent treatment. Mike's crew seemed to forget how to do the most basic activities. The amount of rework for the crew seemed to be increasing. By the end of the month, Mike's crew was behind schedule.

1. How do you think Mike should have handled the confrontation with Ron and Steve?

_____

_____

2. What do you suggest Mike could do about the silent treatment he got from Steve and Ron?

_____

_____

3. If you were Mike, what would you do to get your crew back on schedule?

_____

_____

4. What acts of leadership could be used to get the crew's willing cooperation?

_____

_____

5. To which leadership traits do they correspond?

_____

_____

1. A crew leader differs from a craftworker in that a _____.

   a. crew leader need not have direct experience in those job duties that a craftworker typically performs
   b. crew leader can expect to oversee one or more craftworkers in addition to performing some of the typical duties of the craftworker
   c. crew leader is exclusively in charge of overseeing, since performing technical work is not part of this role
   d. crew leader's responsibilities do not include being present on the job site

2. Among the many traits effective leaders should have is _____.

   a. the ability to communicate the goals of a project
   b. the drive necessary to carry the workload by themselves in order to achieve a goal
   c. a perfectionist nature, which ensures that they will not make useless mistakes
   d. the ability to make decisions without needing to listen to the opinions of others

3. Of the three styles of leadership, the _____ style would be effective in dealing with a craftworker's negative attitude.

   a. facilitator
   b. commander
   c. partner
   d. dictator

4. One way to overcome barriers to effective communication is to _____.

   a. avoid taking notes on the content of the message, since this can be distracting
   b. avoid reacting emotionally to the message
   c. anticipate the content of the message and interrupt if necessary in order to show interest
   d. think about how to respond to the message while listening

5. Feedback is important in verbal communication because it requires the _____.

   a. sender to repeat the message
   b. receiver to restate the message
   c. sender to avoid technical jargon
   d. sender to concentrate on the message

6. A good way to motivate employees is to use a one-size-fits-all approach, since employees are members of a team with a common goal.

   a. True
   b. False

7. A crew leader can effectively delegate responsibilities by _____.

   a. refraining from evaluating the employee's performance once the task is completed, since it is a new task for the employee
   b. doing the job for the employee to make sure the task is done correctly
   c. allowing the employee to give feedback and suggestions about the task
   d. communicating information to the employee, generally in written form

8. Problem solving differs from decision making in that _____.

   a. problem solving involves identifying discrepancies between the way a situation is and the way it should be
   b. decision making involves separating facts from non-facts
   c. decision making involves eliminating differences
   d. problem solving involves determining an alternative course of action for a given situation

Objectives

Upon completion of this section, you will be able to:

1. Explain the importance of safety.
2. Give examples of direct and indirect costs of workplace accidents.
3. Identify safety hazards of the construction industry.
4. Explain the purpose of OSHA.
5. Discuss OSHA inspection procedures.
6. Identify the key points of a safety program.
7. List steps to train employees on how to perform new tasks safely.
8. Identify a crew leader's safety responsibilities.
9. Explain the importance of having employees trained in first aid and cardiopulmonary resuscitation (CPR).
10. Describe the indications of substance abuse.
11. List the essential parts of an accident investigation.
12. Describe ways to maintain employee interest in safety. Distinguish between company policies and procedures.

1.0.0 SAFETY OVERVIEW

Businesses lose millions of dollars every year because of on-the-job accidents. Work-related injuries, sickness, and deaths have caused untold suffering for workers and their families. Project delays and budget overruns from injuries and fatalities result in huge losses for employers, and work-site accidents erode the overall morale of the crew.

Craftworkers are exposed to hazards as part of the job. Examples of these hazards include falls from heights, working on scaffolds, using cranes in the presence of power lines, operating heavy machinery, and working on electrically-charged or pressurized equipment. Despite these hazards, experts believe that applying preventive safety measures could drastically reduce the number of accidents.

As a crew leader, one of your most important tasks is to enforce the company's safety program and make sure that all workers are performing their tasks safely. To be successful, the crew leader should:

- Be aware of the costs of accidents.
- Understand all federal, state, and local governmental safety regulations.
- Be involved in training workers in safe work methods.
- Conduct training sessions.
- Get involved in safety inspections, accident investigations, and fire protection and prevention.

Crew leaders are in the best position to ensure that all jobs are performed safely by their crew members. Providing employees with a safe working environment by preventing accidents and enforcing safety standards will go a long way towards maintaining the job schedule and enabling a job's completion on time and within budget.

1.1.0 Accident Statistics

Each day, workers in construction and industrial occupations face the risk of falls, machinery accidents, electrocutions, and other potentially fatal occupational hazards.

The National Institute of Occupational Safety and Health (NIOSH) statistics show that about 1,000 construction workers are killed on the job each year, more fatalities than in any other industry. Falls are the leading cause of deaths in the construction industry, accounting for over 40 percent of the fatalities. Nearly half of the fatal falls occurred from roofs, scaffolds, or ladders. Roofers, structural metal workers, and painters experienced the greatest number of fall fatalities.

In addition to the number of fatalities that occur each year, there are a staggering number of work-related injuries. In 2007, for example, more than 135,000 job-related injuries occurred in the construction industry. NIOSH reports that approximately 15 percent of all worker's compensation costs are spent on injured construction workers. The causes of injuries on construction sites include falls, coming into contact with electric current, fires, and mishandling of machinery or equipment. According to NIOSH, back injuries are the leading safety problem in workplaces.

Did you know?

When OSHA inspects a job site, they focus on the types of safety hazards that are most likely to cause fatal injuries. These hazards fall into the following classifications:

- Falls from elevations
- Struck-by hazards
- Caught in/between hazards
- Electrical shock hazards

2.0.0 Costs of Accidents

Occupational accidents are estimated to cost more than $100 billion every year. These costs affect the employee, the company, and the construction industry as a whole.

Organizations encounter both direct and indirect costs associated with workplace accidents. Examples of direct costs include workers' compensation claims and sick pay; indirect costs include increased absenteeism, loss of productivity, loss of job opportunities due to poor safety records, and negative employee morale attributed to workplace injuries. There are many other related costs involved with workplace accidents. A company can be insured against some of them, but not others. To compete and survive, companies must control these as well as all other employment-related costs.

2.1.0 Insured Costs

Insured costs are those costs either paid directly or reimbursed by insurance carriers. Insured costs related to injuries or deaths include the following:

- Compensation for lost earnings (known as worker's comp)
- Medical and hospital costs
- Monetary awards for permanent disabilities
- Rehabilitation costs
- Funeral charges
- Pensions for dependents

Insurance premiums or charges related to property damages include:

- Fire
- Loss and damage
- Use and occupancy
- Public liability
- Replacement cost of equipment, material, and structures

2.2.0 Uninsured Costs

The costs related to accidents can be compared to an iceberg, as shown in *Figure 7*. The tip of the iceberg represents direct costs, which are the visible costs. The more numerous indirect costs are not readily measureable, but they can represent a greater burden than the direct costs.

Uninsured costs related to injuries or deaths include the following:

- First aid expenses
- Transportation costs
- Costs of investigations
- Costs of processing reports
- Down time on the job site
- Costs to train replacement workers

Uninsured costs related to wage losses include:

- Idle time of workers whose work is interrupted
- Time spent cleaning the accident area
- Time spent repairing damaged equipment
- Time lost by workers receiving first aid
- Costs of training injured workers in a new career

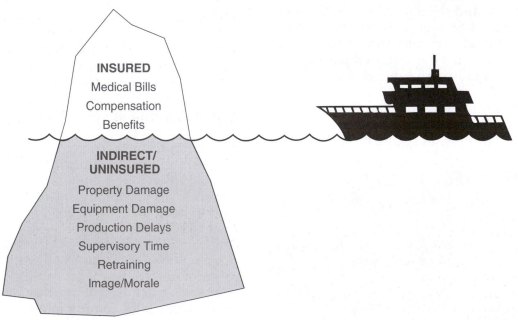

46101-11_F07.EPS

Figure 7 Costs associated with accidents.

Uninsured costs related to production losses include:

- Product spoiled by accident
- Loss of skill and experience
- Lowered production or worker replacement
- Idle machine time

Associated costs may include the following:

- Difference between actual losses and amount recovered
- Costs of rental equipment used to replace damaged equipment
- Costs of new workers used to replace injured workers
- Wages or other benefits paid to disabled workers
- Overhead costs while production is stopped
- Impact on schedule
- Loss of bonus or payment of forfeiture for delays

Uninsured costs related to off-the-job activities include:

- Time spent on injured workers' welfare
- Loss of skill and experience of injured workers
- Costs of training replacement workers

Uninsured costs related to intangibles include:

- Lowered employee morale
- Increased labor conflict
- Unfavorable public relations
- Loss of bid opportunities because of poor safety records
- Loss of client goodwill

3.0.0 Safety Regulations

To reduce safety and health risks and the number of injuries and fatalities on the job, the federal government has enacted laws and regulations, including the *Occupational Safety and Health Act of 1970*. The purpose of OSHA is "to assure so far as possible every working man and woman in the Nation safe and healthful working conditions and to preserve our human resources."

To promote a safe and healthy work environment, OSHA issues standards and rules for working conditions, facilities, equipment, tools, and work processes. It does extensive research into occupational accidents, illnesses, injuries, and deaths in an effort to reduce the number of occurrences and adverse effects. In addition, OSHA regulatory agencies conduct workplace inspections to ensure that companies follow the standards and rules.

3.1.0 Workplace Inspections

To enforce OSHA regulations, the government has granted regulatory agencies the right to enter public and private properties to conduct workplace safety investigations. The agencies also have the right to take legal action if companies are not in compliance with the Act. These regulatory agencies employ OSHA Compliance Safety and Health Officers (CSHOs), who are chosen for their knowledge in the occupational safety and health field. The CSHOs are thoroughly trained in OSHA standards and in recognizing safety and health hazards.

States with their own occupational safety and health programs conduct inspections. To do so, they enlist the services of qualified state CSHOs.

Companies are inspected for a multitude of reasons. They may be randomly selected, or they may be chosen because of employee complaints, due to an imminent danger, or as a result of major accidents or fatalities.

OSHA can assess significant financial penalties for safety violations. In some cases, a superintendent or crew leader can be held criminally liable for repeat violations.

3.2.0 Penalties for Violations

OSHA has established monetary fines for the violation of their regulations. The penalties as of 2010 are shown in *Table 1*.

In addition to the fines, there are possible criminal charges for willful violations resulting in death or serious injury. There can also be personal liability for failure to comply with OSHA regulations. The attitude of the employer and their safety history can have a significant effect on the outcome of a case.

Table 1 OSHA Penalties for Violations

Violation	Penalty
Willful Violations	Maximum $70,000
Repeated Violations	Minimum $70,000
Serious, Other-than-Serious, Other Specific Violations	Minimum $7,000
OSHA Notice Violation	$1,000
Failure to Post *OSHA 300A Summary of Work-Related Injuries and Illnesses*	$1,000
Failure to Properly Maintain *OSHA 300 Log of Work-Related Injuries and Illnesses*	$1,000
Failure to Promptly and Properly Report Fatality/Catastrophe	$5,000
Failure to Permit Access to Records Under *OSHA 1904* Regulations	$1,000
Failure to Follow Advance Notification Requirements Under *OSHA 1903.6* Regulations	$2,000
Failure to Abate – for Each Calendar Day Beyond Abatement Date	$7,000
Retaliation Against Individual for Filing OSHA Complaint	$10,000

Did you know?

Nearly half the states in the U.S. have state-run OSHA programs. These programs are set up under federal OSHA guidelines and must establish job health and safety standards that are at least as effective as the federal standards. The states have the option of adopting more stringent standards or setting standards for hazards not addressed in the federal program. Of the 22 states with state-run OSHA programs, eight of them limit their coverage to public employees.

4.0.0 EMPLOYER SAFETY RESPONSIBILITIES

Each employer must set up a safety and health program to manage workplace safety and health and to reduce work-related injuries, illnesses, and fatalities. The program must be appropriate for the conditions of the workplace. It should consider the number of workers employed and the hazards to which they are exposed while at work.

To be successful, the safety and health program must have management leadership and employee participation. In addition, training and informational meetings play an important part in effective programs. Being consistent with safety policies is the key. Regardless of the employer's responsibility, however, the individual worker is ultimately responsible for his or her own safety.

4.1.0 Safety Program

The crew leader plays a key role in the successful implementation of the safety program. The crew leader's attitudes toward the program set the standard for how crew members view safety. Therefore, the crew leader should follow all program guidelines and require crew members to do the same.

Safety programs should consist of the following:

- Safety policies and procedures
- Hazard identification and assessment
- Safety information and training
- Safety record system
- Accident investigation procedures
- Appropriate discipline for not following safety procedures
- Posting of safety notices

4.1.1 Safety Policies and Procedures

Employers are responsible for following OSHA and state safety standards. Usually, they incorporate OSHA and state regulations into a safety policies and procedures manual. Such a manual is presented to employees when they are hired.

Basic safety requirements should be presented to new employees during their orientation to the company. If the company has a safety manual, the new employee should be required to read it and sign a statement indicating that it is understood. If the employee cannot read, the employer should have someone read it to the employee and answer

---

any questions that arise. The employee should then sign a form stating that he or she understands the information.

It is not enough to tell employees about safety policies and procedures on the day they are hired and never mention them again. Rather, crew leaders should constantly emphasize and reinforce the importance of following all safety policies and procedures. In addition, employees should play an active role in determining job safety hazards and find ways that the hazards can be prevented and controlled.

4.1.2 Hazard Identification and Assessment

Safety policies and procedures should be specific to the company. They should clearly present the hazards of the job. Therefore, crew leaders should also identify and assess hazards to which employees are exposed. They must also assess compliance with OSHA and state standards.

To identify and assess hazards, OSHA recommends that employers conduct inspections of the workplace, monitor safety and health information logs, and evaluate new equipment, materials, and processes for potential hazards before they are used.

Crew leaders and employees play important roles in identifying hazards. It is the crew leader's responsibility to determine what working conditions are unsafe and to inform employees of hazards and their locations. In addition, they should encourage their crew members to tell them about hazardous conditions. To accomplish this, crew leaders must be present and available on the job site.

The crew leader also needs to help the employee be aware of and avoid the built-in hazards to which craftworkers are exposed. Examples include working at elevations, working in confined spaces such as tunnels and underground vaults, on caissons, in excavations with earthen walls, and other naturally dangerous projects.

In addition, the crew leader can take safety measures, such as installing protective railings to prevent workers from falling from buildings, as well as scaffolds, platforms, and shoring.

4.1.3 Safety Information and Training

The employer must provide periodic information and training to new and long-term employees. This must be done as often as necessary so that all employees are adequately trained. Special training and informational sessions must be provided when safety and health information changes or workplace conditions create new hazards. It is important to note that safety-related information must be presented in a manner that the employee will understand.

Whenever a crew leader assigns an experienced employee a new task, the crew leader must ensure that the employee is capable of doing the work in a safe manner. The crew leader can accomplish this by providing safety information or training for groups or individuals.

The crew leader should do the following:

- Define the task.
- Explain how to do the task safely.
- Explain what tools and equipment to use and how to use them safely.
- Identify the necessary personal protective equipment.
- Explain the nature of the hazards in the work and how to recognize them.
- Stress the importance of personal safety and the safety of others.
- Hold regular safety training sessions with the crew's input.
- Review material safety data sheets (MSDSs) that may be applicable.

4.1.4 Safety Record Systems

OSHA regulations (*CFR 29, Part 1904*) require that employers keep records of hazards identified and document the severity of the hazard. The information should include the likelihood of employee exposure to the hazard, the seriousness of the harm associated with the hazard, and the number of exposed employees.

In addition, the employer must document the actions taken or plans for action to control the hazards. While it is best to take corrective action immediately, it is sometimes necessary to develop a plan for the purpose of setting priorities and deadlines and tracking progress in controlling hazards.

Employers who are subject to the recordkeeping requirements of the *Occupational Safety and Health Act of 1970* must maintain a log of all recordable occupational injuries and illnesses. This is known as the *OSHA 300/300A* form.

An MSDS is designed to provide both workers and emergency personnel with the proper procedures for handling or working with a substance that may be dangerous. An MSDS will include information such as physical data (melting point, boiling point, flash point, etc.), toxicity, health effects, first aid, reactivity, storage, disposal, protective equipment, and spill/leak procedures. These sheets are of particular use if a spill or other accident occurs.

Any company with 11 or more employees must post an *OSHA 300A* form, *Log of Work-Related Injuries and Illnesses,* between February 1 and April 30 of each year. Employees have the right to review this form. Check your company's policies with regard to these reports.

OSHA's Form 300A (Rev. 01/2004)

Summary of Work-Related Injuries and Illnesses

Year 20___

U.S. Department of Labor
Occupational Safety and Health Administration

Form approved OMB no. 1218-0176

All establishments covered by Part 1904 must complete this Summary page, even if no work-related injuries or illnesses occurred during the year. Remember to review the Log to verify that the entries are complete and accurate before completing this summary.

Using the Log, count the individual entries you made for each category. Then write the totals below, making sure you've added the entries from every page of the Log. If you had no cases, write "0."

Employees, former employees, and their representatives have the right to review the OSHA Form 300 in its entirety. They also have limited access to the OSHA Form 301 or its equivalent. See 29 CFR Part 1904.35, in OSHA's recordkeeping rule, for further details on the access provisions for these forms.

Number of Cases

Total number of deaths

(G)

Total number of cases with days away from work

(H)

Total number of cases with job transfer or restriction

(I)

Total number of other recordable cases

(J)

Number of Days

Total number of days away from work

(K)

Total number of days of job transfer or restriction

(L)

Injury and Illness Types

Total number of . . .
(M)

(1) Injuries

(2) Skin disorders

(3) Respiratory conditions

(4) Poisonings

(5) Hearing loss

(6) All other illnesses

Post this Summary page from February 1 to April 30 of the year following the year covered by the form.

Public reporting burden for this collection of information is estimated to average 58 minutes per response, including time to review the instructions, search and gather the data needed, and complete and review the collection of information. Persons are not required to respond to the collection of information unless it displays a currently valid OMB control number. If you have any comments about these estimates or any other aspects of this data collection, contact: US Department of Labor, OSHA Office of Statistical Analysis, Room N-3644, 200 Constitution Avenue, NW, Washington, DC 20210. Do not send the completed forms to this office.

Establishment information

Your establishment name

Street

City _____ State ____ ZIP ____

Industry description (e.g., Manufacture of motor truck trailers)

Standard Industrial Classification (SIC), if known (e.g., 3715)

OR

North American Industrial Classification (NAICS), if known (e.g., 336212)

Employment information (If you don't have these figures, see the Worksheet on the back of this page to estimate.)

Annual average number of employees

Total hours worked by all employees last year

Sign here

Knowingly falsifying this document may result in a fine.

I certify that I have examined this document and that to the best of my knowledge the entries are true, accurate, and complete.

Company executive _____ Title _____

Phone ( ) _____ Date / /

46101-11_SA01.EPS

Logs must be maintained and retained for five years following the end of the calendar year to which they relate. Logs must be available (normally at the establishment) for inspection and copying by representatives of the Department of Labor, the Department of Health and Human Services, or states accorded jurisdiction under the Act. Employees, former employees, and their representatives may also have access to these logs.

4.1.5 Accident Investigation

In the event of an accident, the employer is required to investigate the cause of the accident and determine how to avoid it in the future. According to OSHA, the employer must investigate each work-related death, serious injury or illness, or incident having the potential to cause death or serious physical harm. The employer should document any findings from the investigation, as well as the action plan to prevent future occurrences. This should be done immediately, with photos or video if possible. It is important that the investigation uncover the root cause of the accident so that it can be avoided in the future. In many cases, the root cause can be traced to a flaw in the system that failed to recognize the unsafe condition or the potential for an unsafe act (*Figure 8*).

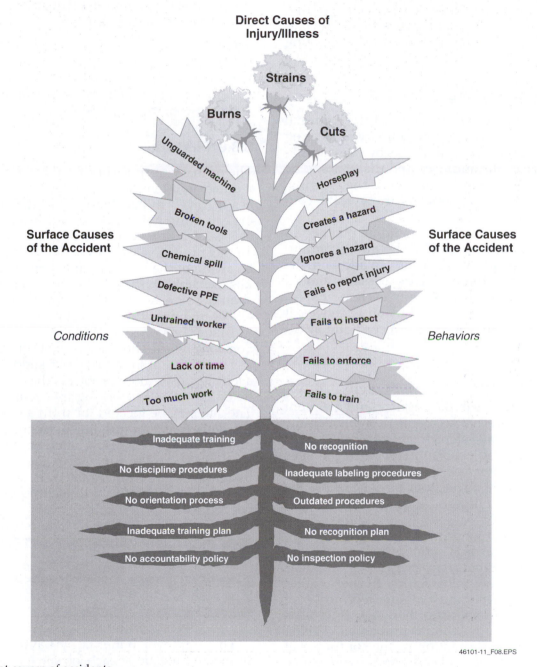

Figure 8 Root causes of accidents.

5.0.0 CREW LEADER INVOLVEMENT IN SAFETY

To be effective leaders, crew leaders must be actively involved in the safety program. Crew leader involvement includes conducting frequent safety training sessions and inspections; promoting first aid and fire protection and prevention; preventing substance abuse on the job; and investigating accidents.

5.1.0 Safety Training Sessions

A safety training session may be a brief, informal gathering of a few employees or a formal meeting with instructional films and talks by guest speakers. The size of the audience and the topics to be addressed determine the format of the meeting. Small, informal safety sessions are typically conducted weekly.

Safety training sessions should be planned in advance, and the information should be communicated to all employees affected. In addition, the topics covered in these training sessions should be timely and practical. A log of each safety session must be kept and signed by all attendees. It must then be maintained as a record and available for inspection.

5.2.0 Inspections

Crew leaders must make regular and frequent inspections to prevent accidents from happening. They must also take steps to avoid accidents. For that purpose, they need to inspect the job sites where their workers perform tasks. It is recommended that these inspections be done before the start of work each day and during the day at different times.

Crew leaders must protect workers from existing or potential hazards in their work areas. Crew leaders are sometimes required to work in areas controlled by other contractors. In these situations, the crew leader must maintain control over the safety exposure of his or her crew. If hazards exist, the crew leader should immediately bring the hazards to the attention of the contractor at fault, their superior, and the person responsible for the job site.

Crew leader inspections are only valuable if action is taken to correct potential hazards. Therefore, crew leaders must be alert for unsafe acts on their work sites. When an employee performs an unsafe action, the crew leader must explain to the employee why the act was unsafe, ask that the employee not do it again, and request cooperation in promoting a safe working environment. The crew leader must document what happened and what the employee was asked to do to correct the situation. It is then very important that crew leaders follow up to make certain the employee is complying with the safety procedures. Never allow a safety violation to go uncorrected. There are three courses of action that you, as a crew leader, can take in an unsafe situation:

- Get the appropriate party to correct the problem.
- Fix the problem yourself.
- Refuse to have the crew work in the area until the problem is corrected.

5.3.0 First Aid

The primary purpose of first aid is to provide immediate and temporary medical care to employees involved in accidents, as well as employees experiencing non-work-related health emergencies, such as chest pains or breathing difficulty. To meet this objective, every crew leader should be aware of the location and contents of first aid kits available on the job site. Emergency numbers should be posted in the job trailer. In addition, OSHA requires that at least one person trained in first aid be present at the job site at all times. Someone on site should also be trained in CPR.

The victim of an accident or sudden illness at a job site poses more problems than normal since he or she may be working in a remote location. The site may be located far from a rescue squad, fire department, or hospital, presenting a problem in the rescue and transportation of the victim to a hospital. The worker may also have been injured by falling rock or other materials, so special rescue equipment or first-aid techniques are often needed.

NOTE

CPR training must be renewed every two years.

The employer benefits by having personnel trained in first aid at each job site in the following ways:

- The immediate and proper treatment of minor injuries may prevent them from developing into more serious conditions. As a result, medical expenses, lost work time, and sick pay may be eliminated or reduced.
- It may be possible to determine if professional medical attention is needed.
- Valuable time can be saved when a trained individual prepares the patient for treatment when professional medical care arrives. As a result, lives can be saved.

The American Red Cross, Medic First Aid, and the United States Bureau of Mines provide basic and advanced first aid courses at nominal costs. These courses include both first aid and CPR. The local area offices of these organizations can provide further details regarding the training available.

5.4.0 Fire Protection and Prevention

Fires and explosions kill and injure many workers each year, so it is important that crew leaders understand and practice fire-prevention techniques as required by company policy.

The need for protection and prevention is increasing as new building materials are introduced. Some building materials are highly flammable. They produce great amounts of smoke and gases, which cause difficulties for fire fighters, and can quickly overcome anyone present. Other materials melt when ignited and may spread over floors, preventing fire-fighting personnel from entering areas where this occurs.

OSHA has specific standards for fire safety. They require that employers provide proper exits, fire-fighting equipment, and employee training on fire prevention and safety. For more information, consult OSHA guidelines.

5.5.0 Substance Abuse

Unfortunately, drug and alcohol abuse is a continuing problem in the workplace. Drug abuse means inappropriately using drugs, whether they are legal or illegal. Some people use illegal street drugs, such as cocaine or marijuana. Others use legal prescription drugs incorrectly by taking too many pills, using other people's medications, or self-medicating. Others consume alcohol to the point of intoxication.

It is essential that crew leaders enforce company policies and procedures regarding substance abuse. Crew leaders must work with management to deal with suspected drug and alcohol abuse and should not handle these situations themselves. These cases are normally handled by the Human Resources Department or designated manager. There are legal consequences of drug and alcohol abuse and the associated safety implications. If you suspect that an employee is suffering from drug or alcohol abuse, immediately contact your supervisor and/or Human Resources Department for assistance. That way, the business and the employee's safety are protected.

It is the crew leader's responsibility to make sure that safety is maintained at all times. This may include removing workers from a work site where they may be endangering themselves or others.

For example, suppose several crew members go out and smoke marijuana or have a few beers during lunch. Then, they return to work to erect scaffolding for a concrete pour in the afternoon. If you can smell marijuana on the crew member's clothing or alcohol on their breath, you must step in and take action. Otherwise, they might cause an accident that could delay the project or cause serious injury or death to themselves or others.

It is often difficult to detect drug and alcohol abuse because the effects can be subtle. The best way is to look for identifiable effects, such as those mentioned above or sudden changes in behavior that are not typical of the employee. Some examples of such behaviors include the following:

- Unscheduled absences; failure to report to work on time
- Significant changes in the quality of work
- Unusual activity or lethargy
- Sudden and irrational temper flare-ups
- Significant changes in personal appearance, cleanliness, or health

There are other more specific signs that should arouse suspicion, especially if more than one is exhibited. Among them are:

- Slurring of speech or an inability to communicate effectively
- Shiftiness or sneaky behavior, such as an employee disappearing to wooded areas, storage areas, or other private locations
- Wearing sunglasses indoors or on overcast days to hide dilated or constricted pupils
- Wearing long-sleeved garments, particularly on hot days, to cover marks from needles used to inject drugs
- Attempting to borrow money from co-workers
- The loss of an employee's tools and company equipment

5.6.0 Job-Related Accident Investigations

There are two times when a crew leader may be involved with an accident investigation. The first time is when an accident, injury, or report of work-connected illness takes place. If present on site, the crew leader should proceed immediately to the accident location to ensure that proper first aid is being provided. He or she will also want to make sure that other safety and operational measures are taken to prevent another incident.

If mandated by company policy, the crew leader will need to make a formal investigation and submit a report after an incident. An investigation looks for the causes of the accident by examining the situation under which it occurred and talking to the people involved. Investigations are perhaps the most useful tool in the prevention of future accidents.

There are four major parts to an accident investigation. The crew leader will be concerned with each one. They are:

- Describing the accident
- Determining the cause of the accident
- Determining the parties involved and the part played by each
- Determining how to prevent re-occurrences

Case Study

For years, a prominent safety engineer was confused as to why sheet metal workers fractured their toes frequently. The crew leader had not performed thorough accident investigations, and the injured workers were embarrassed to admit how the accidents really occurred. It was later discovered they used the metal reinforced cap on their safety shoes as a "third hand" to hold the sheet metal vertically in place when they fastened it. The sheet metal was inclined to slip and fall behind the safety cap onto the toes, causing fractures. Several injuries could have been prevented by performing a proper investigation after the first accident.

6.0.0 PROMOTING SAFETY

The best way for crew leaders to encourage safety is through example. Crew leaders should be aware that their behavior sets standards for their crew members. If a crew leader cuts corners on safety, then the crew members may think that it is okay to do so as well.

The key to effectively promote safety is good communication. It is important to plan and coordinate activities and to follow through with safety programs. The most successful safety promotions occur when employees actively participate in planning and carrying out activities.

Some activities used by organizations to help motivate employees on safety and help promote safety awareness include:

- Safety training sessions
- Contests
- Recognition and awards
- Publicity

6.1.0 Safety Training Sessions

Safety training sessions can help keep workers focused on safety and give them the opportunity to discuss safety concerns with the crew. This topic was addressed in a previous section.

6.2.0 Safety Contests

Contests are a great way to promote safety in the workplace. Examples of safety-related contests include the following:

- Sponsoring housekeeping contests for the cleanest job site or work area
- Challenging employees to come up with a safety slogan for the company or department
- Having a poster contest that involves employees or their children creating safety-related posters
- Recording the number of accident-free workdays or person-hours
- Giving safety awards (hats, T-shirts, other promotional items or prizes)

One of the positive aspects of contests is their ability to encourage employee participation. It is important, however, to ensure that the contest has a valid purpose. For example, the posters or slogans created in a poster contest can be displayed throughout the organization as safety reminders.

6.3.0 Incentives and Awards

Incentives and awards serve several purposes. Among them are acknowledging and encouraging good performance, building goodwill, reminding employees of safety issues, and publicizing the importance of practicing safety standards. There are countless ways to recognize and award safety. Examples include the following:

- Supplying food at the job site when a certain goal is achieved
- Providing a reserved parking space to acknowledge someone for a special achievement
- Giving gift items such as T-shirts or gift certificates to reward employees
- Giving plaques to a department or an individual (*Figure 9*)
- Sending a letter of appreciation
- Publicly honoring a department or an individual for a job well done

Creativity can be used to determine how to recognize and award good safety on the work site. The only precautionary measure is that the award should be meaningful and not perceived as a bribe. It should be representative of the accomplishment.

6.4.0 Publicity

Publicizing safety is the best way to get the message out to employees. An important aspect of publicity is to keep the message accurate and current. Safety posters that are hung for years on end tend to lose effectiveness. It is important to keep ideas fresh.

Examples of promotional activities include posters or banners, advertisements or information on bulletin boards, payroll mailing stuffers, and employee newsletters. In addition, merchandise can be purchased that promotes safety, including buttons, hats, T-shirts, and mugs.

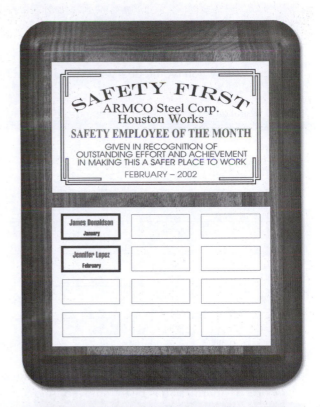

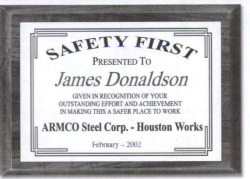

46101-11_F09.EPS

Figure 9 Examples of safety plaques.

Described here are three scenarios that reflect unsafe practices by craft workers. For each of these scenarios write down how you would deal with the situation, first as the crew leader of the craft worker, and then as the leader of another crew.

1. You observe a worker wearing his hard hat backwards and his safety glasses hanging around his neck. He is using a concrete saw.

_____

_____

_____

_____

_____

_____

_____

2. As you are supervising your crew on the roof deck of a building under construction, you notice that a section of guard rail has been removed. Another contractor was responsible for installing the guard rail.

_____

_____

_____

_____

_____

_____

_____

3. Your crew is part of plant shutdown at a power station. You observe that a worker is welding without a welding screen in an area where there are other workers.

_____

_____

_____

_____

_____

_____

_____

Review Questions

1. One of a crew leader's responsibilities is to enforce company safety policies.

   a. True
   b. False

2. Which of the following is an indirect cost of an accident?

   a. Medical bills
   b. Production delays
   c. Compensation
   d. Employee benefits

3. A crew leader can be held criminally liable for repeat safety violations.

   a. True
   b. False

4. OSHA inspection of a business or job site _____.

   a. can be done only by invitation
   b. is done only after an accident
   c. can be conducted at random
   d. can be conducted only if a safety violation occurs

5. The *OSHA 300* form deals with _____.

   a. penalties for safety violations
   b. workplace illnesses and injuries
   c. hazardous material spills
   d. safety training sessions

6. A crew leader's responsibilities include all of the following, *except* _____.

   a. conducting safety training sessions
   b. developing a company safety program
   c. performing safety inspections
   d. participating in accident investigations

7. In order to ensure workplace safety, the crew leader should _____.

   a. hold formal safety training sessions
   b. have crew members conduct on-site safety inspections
   c. notify contractors and their supervisor of hazards in a situation where a job is being performed in an unsafe area controlled by other contractors
   d. hold crew members responsible for making a formal report and investigation following an accident

8. Prohibitions on the abuse of drugs deals only with illegal drugs such as cocaine and marijuana.

   a. True
   b. False

SECTION FOUR
PROJECT CONTROL

Objectives

Upon completion of this section, you will be able to:

1. Describe the three phases of a construction project.
2. Define the three types of project delivery systems.
3. Define planning and describe what it involves.
4. Explain why it is important to plan.
5. Describe the two major stages of planning.
6. Explain the importance of documenting job site work.
7. Describe the estimating process.
8. Explain how schedules are developed and used.
9. Identify the two most common schedules.
10. Explain how the critical path method (CPM) of scheduling is used.
11. Describe the different costs associated with building a job.
12. Explain the crew leader's role in controlling costs.
13. Illustrate how to control the main resources of a job: materials, tools, equipment, and labor.
14. Explain the differences between production and productivity and the importance of each.

Performance Tasks

1. Develop and present a look-ahead schedule.
2. Develop an estimate for a given work activity

1.0.0 PROJECT CONTROL OVERVIEW

The contractor, project manager, superintendent, and crew leader each have management responsibilities for their assigned jobs. For example, the contractor's responsibility begins with obtaining the contract, and it does not end until the client takes ownership of the project. The project manager is generally the person with overall responsibility for coordinating the project. Finally, the superintendent and crew leader are responsible for coordinating the work of one or more workers, one or more crews of workers within the company and, on occasion, one or more crews of subcontractors. The crew leader directs a crew in the performance of work tasks.

This section describes methods of effective and efficient project control. It examines estimating, planning and scheduling, and resource and cost control. All the workers who participate in the job are responsible at some level for controlling cost and schedule performance and for ensuring that the project is completed according to plans and specifications.

> **NOTE**
>
> The material in this section is based largely on building-construction projects. However, the project control principles described here apply generally to all types of projects.

Construction projects are made up of three phases: the development phase, the planning phase, and the construction phase.

1.1.0 Development Phase

A new building project begins when an owner has decided to build a new facility or add to an existing facility. The development process is the first stage of planning for a new building project. This process involves land research and feasibility studies to ensure that the project has merit. Architects or engineers develop the conceptual drawings that define the project graphically. They then provide the owner with sketches of room layouts and elevations and make suggestions about what construction materials should be used.

During the development phase, an estimate for the proposed project is developed in order to establish a preliminary budget. Once that budget has been established, the financing of the project is discussed with lending institutions. The architects/engineers and/or the owner begins preliminary reviews with government agencies. These reviews include zoning, building restrictions, landscape requirements, and environmental impact studies.

Also during the development phase, the owner must analyze the project's cost and potential return on investment (ROI) to ensure that its costs will not exceed its market value and that the project provides a good return on investment. If the project passes this test, the architect/engineer will proceed to the planning phase.

1.2.0 Planning Phase

When the architect/engineer begins to develop the project drawings and specifications, other design professionals such as structural, mechanical, and electrical engineers are brought in. They perform the calculations, make a detailed technical analysis, and check details of the project for accuracy.

The design professionals create drawings and specifications. These drawings and specifications are used to communicate the necessary information to the contractors, subcontractors, suppliers, and workers that contribute to a project.

During the planning phase, the owners hold many meetings to refine estimates, adjust plans to conform to regulations, and secure a construction loan. If the project is a condominium, an office building, or a shopping center, then a marketing program must be developed. In such cases, the selling of the project is often started before actual construction begins.

Next, a complete set of drawings, specifications, and bid documents is produced. Then the owner will select the method to obtain contractors. The owner may choose to negotiate with several contractors or select one through competitive bidding. Note that safety must also be considered as part of the planning process. A safety crew leader may walk through the site as part of the pre-bid process.

Contracts can take many forms. The three basic types from which all other types are derived are firm fixed price, cost reimbursable, and guaranteed maximum price.

- *Firm fixed price* – In this type of contract, the buyer generally provides detailed drawings and specifications, which the contractor uses to calculate the cost of materials and labor. To these costs, the contractor adds a percentage representing company overhead expenses such as office rent, insurance, and accounting/payroll costs. On top of all this, the contractor adds a profit factor. When submitting the bid, the contractor will state very specifically the conditions and assumptions on which the bid is based. These conditions and assumptions form the basis from which changes can be priced. Because the price is established in advance, any changes in the job requirements once the job is started will impact the contractor's profit margin. This is where the crew leader can play an important role by identifying problems that increase the amount of labor or material that was planned. By passing this information up the chain of command, the crew leader allows the company to determine if the change is outside the scope of the bid. If so, they can submit a change order request to cover the added cost.
- *Cost reimbursable* – In this type of contract, the buyer reimburses the contractor for labor, materials, and other costs encountered in the performance of the contract. Typically, the contractor and buyer agree in advance on hourly or daily labor rates for different categories of worker.

These rates include an amount representing the contractor's overhead expense. The buyer also reimburses the contractor for the cost of materials and equipment used on the job. The buyer and contractor also negotiate a profit margin. On this type of contract, the profit margin is likely to be lower than that of a fixed-price contract because the contractor's cost risk is significantly reduced. The profit margin is often subject to incentive or penalty clauses that make the amount of profit awarded subject to performance by the contractor. Performance is usually tied to project schedule milestones.

- *Guaranteed maximum price (GMP)* – This form of contract, also called a not-to-exceed contract, is most often used on projects that have been negotiated with the owner. Involvement in the process usually includes preconstruction, and the entire team develops the parameters that define the basis for the work. In some instances, the owner will require a competitively-bid GMP. In such cases, the scope of work has not been fully defined, but bids are taken for general conditions (direct costs) and fee based on an assumed volume of work. The advantages of the GMP contract vehicle are:
  – Reduced design time
  – Allows for phased construction
  – Uses a team approach to a project
  – Reduction in changes related to incomplete drawings

1.3.0 Construction Phase

The designated contractor enlists the help of mechanical, electrical, elevator, and other specialty subcontractors to complete the construction phase. The contractor may perform one or more parts of the construction, and rely on subcontractors for the remainder of the work. However, the general contractor is responsible for managing all the trades necessary to complete the project.

As construction nears completion, the architect/engineer, owner, and government agencies start their final inspections and acceptance of the project. If the project has been managed by the general contractor, the subcontractors have performed their work, and the architect/ engineers have regularly inspected the project to ensure that local codes have been followed, then the inspection procedure can be completed in a timely manner. This results in a satisfied client and a profitable project for all.

On the other hand, if the inspection reveals faulty workmanship, poor design or use of materials, or violation of codes, then the inspection and acceptance will become a lengthy battle and may result in a dissatisfied client and an unprofitable project.

Figure 10 shows the flow of a typical project.

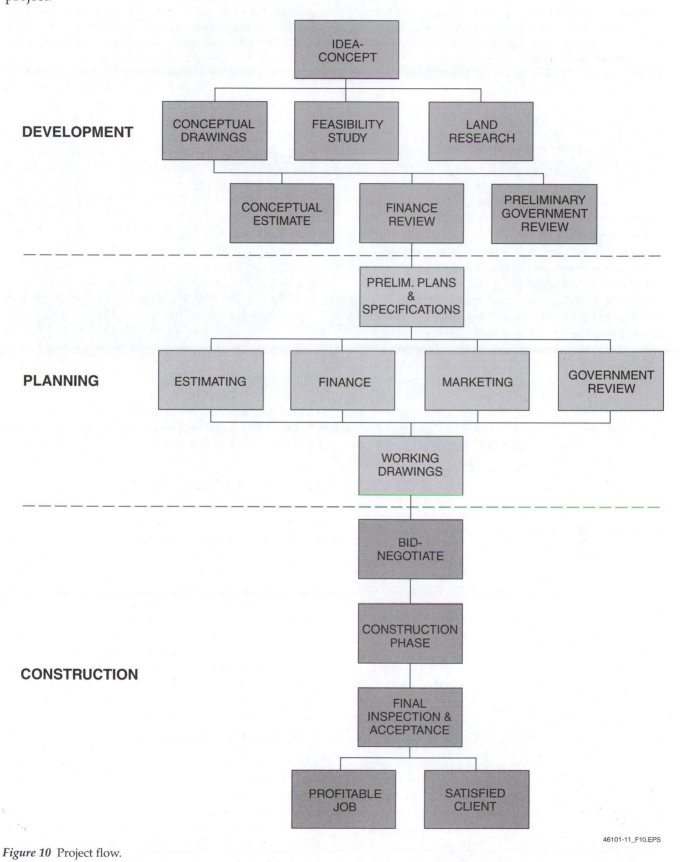

Figure 10 Project flow.

46101-11_F10.EPS

LEED stands for Leadership in Energy and Environmental Design. It is an initiative started by the U.S. Green Building Council (USGBC) to encourage and accelerate the adoption of sustainable construction standards worldwide through a Green Building Rating System™. USGBC is a non-government, not-for-profit group. Their rating system addresses six categories:

1. Sustainable Sites (SS)
2. Water Efficiency (WE)
3. Energy and Atmosphere (EA)
4. Materials and Resources (MR)
5. Indoor Environmental Quality (EQ)
6. Innovation in Design (ID)

LEED is a voluntary program that is driven by building owners. Construction crew leaders may not have input into the decision to seek LEED certification for a project, or what materials are used in the project's construction. However, these crew leaders can help to minimize material waste and support recycling efforts, both of which are factors in obtaining LEED certification.

An important question to ask is whether your project is seeking LEED certification. If the project is seeking certification, the next step is to ask what your role will be in getting the certification. If you are procuring materials, what information is needed and who should receive it? What specifications and requirements do the materials need to meet? If you are working outside the building or inside in a protected area, what do you need to do to protect the work area? How should waste be managed? Are there any other special requirements that will be your responsibility? Do you see any opportunities for improvement? LEED principles are described in more detail in the NCCER publications, *Your Role in the Green Environment* and *Sustainable Construction Supervisor*.

46101-11_SA02.EPS

1.3.1 As-Built Drawings

A set of drawings for a construction project reflects the completed project as conceived by the architect and engineers. During construction, changes usually are necessary because of factors unforeseen during the design phase. For example, if cabling or conduit is re-routed, or equipment is installed in a different location than shown on the original drawing, such changes must be marked on the drawings. Without this record, technicians called to perform maintenance or modify the equipment at a later date will have trouble locating all the cabling and equipment.

Any changes made during construction or installation must be documented on the drawings as the changes are made. Changes are usually made using a colored pen or pencil, so the change can be readily spotted. These drawings are commonly called redlines. When the drawings have been revised to reflect the redline changes, the final drawings are called as-builts, and are so marked. These become the drawings of record for the project.

2.0.0 PROJECT DELIVERY SYSTEMS

Project delivery systems refer to the process by which projects are delivered, from development through construction. Project delivery systems focus on three main systems: general contracting, design-build, and construction management (*Figure 11*).

2.1.0 General Contracting

The traditional project delivery system uses a general contractor. In this type of project, the owner determines the design of the project, and then solicits proposals from general contractors. After selecting a general contractor, the owner contracts directly with that contractor, who builds the project as the prime, or controlling, contractor.

	GENERAL CONTRACTING	DESIGN-BUILD	CONSTRUCTION MANAGEMENT
OWNER	Designs project (or hires architect)	Hires general contractor	Hires construction management company
GENERAL CONTRACTOR	Builds project (with owner's design)	Involved in project design, builds project	Builds, may design (hired by construction management company)
CONSTRUCTION MANAGEMENT COMPANY			Hires and manages general contractor and architect

46101-11_F11.EPS

Figure 11 Project delivery systems.

2.2.0 Design-Build

The design-build project delivery system is different from the general contracting delivery system. In the design-build system, both the design and construction of a project are managed by a single entity. GMP contracts are commonly used in these situations.

2.3.0 Construction Management

The construction management project delivery system uses a construction manager to facilitate the design and construction of a project. Construction managers are very involved in project control; their main concerns are controlling time, cost, and the quality of the project.

3.0.0 COST ESTIMATING AND BUDGETING

Before a project is built, an estimate must be prepared. An estimate is the process of calculating the cost of a project. There are two types of costs to consider, including direct and indirect costs. Direct costs, also known as general conditions, are those that can clearly be assigned to a budget. Indirect costs are overhead costs that are shared by all projects. These costs are generally applied as an overhead percentage to labor and material costs.

Direct costs include the following:

- Materials
- Labor
- Tools
- Equipment

Indirect costs refer to overhead items such as:

- Office rent
- Utilities
- Telecommunications
- Accounting
- Office supplies, signs

The bid price includes the estimated cost of the project as well as the profit. Profit refers to the amount of money that the contractor will make after all of the direct and indirect costs have been paid. If the direct and indirect costs exceed those estimated to perform the job, the difference between the actual and estimated costs must come out of the company's profit. This reduces what the contractor makes on the job.

Profit is the fuel that powers a business. It allows the business to invest in new equipment and facilities, provide training, and to maintain a reserve fund for times when business is slow. In the case of large companies, profitability attracts investors who provide the capital necessary for the business to grow. For these reasons, contractors cannot afford to lose money on a consistent basis. Those who cannot operate profitably are forced out of business. Crew leaders can help their companies remain profitable by managing budget, schedule, quality, and safety adhering to the drawings, specifications, and project schedule.

3.1.0 The Estimating Process

The cost estimate must consider a number of factors. Many companies employ professional cost estimators to do this work. They also maintain performance data for previous projects. This data can be used as a guide in estimating new projects. A complete estimate is developed as follows:

Step 1 Using the drawings and specifications, an estimator records the quantity of the materials needed to construct the job. This is called a quantity takeoff. The information is placed on a takeoff sheet like the one shown in *Figure 12*.

Step 2 Productivity rates are used to estimate the amount of labor required to complete the project. Most companies keep records of these rates for the type and size of the jobs that they perform. The company's estimating department keeps these records.

Step 3 The amount of work to be done is divided by the productivity rate to determine labor hours. For example, if the productivity rate for concrete finishing is 40 square feet per hour, and there are 10,000 square feet of concrete to be finished, then 250 hours of concrete finishing labor is required. This number would be multiplied by the hourly rate for concrete finishing to determine the cost of that labor category. If this work is subcontracted, then the subcontractor's cost estimate, raised by an overhead factor, would be used in place of direct labor cost.

Step 4 The total material quantities are taken from the quantity takeoff sheet and placed on a summary or pricing sheet, an example of which is shown in *Figure 13*. Material prices are obtained from local suppliers, and the total cost of materials is calculated.

Step 5 The cost of equipment needed for the project is determined. This number could reflect rental cost or a factor used by the company when their own equipment is to be used.

Step 6 The total cost of all resources—materials, equipment, tools, and labor—is then totaled on the summary sheet. The unit cost—the total cost divided by the total number of units of material to be put into place—can also be calculated.

Step 7 The cost of taxes, bonds, insurance, subcontractor work, and other indirect costs are added to the direct costs of the materials, equipment, tools, and labor.

Step 8 Direct and indirect costs are summed to obtain the total project cost. The contractor's expected profit is then added to that total.

> **NOTE**
>
> There are software programs available to simplify the cost estimating process. Many of them are tailored to specific industries such as construction and manufacturing, and to specific trades within the industries. For example, there are programs available for electrical and HVAC contractors. Estimating programs are typically set up to include a takeoff form and a form for estimating labor by category. Most of these programs include a data base that contains current prices for labor and materials, so they automatically price the job and produce a bid. Once the job is awarded, the programs can generate purchase orders for materials.

3.1.1 Estimating Material Quantities

The crew leader may be required to estimate quantities of materials.

A set of construction drawings and specifications is needed in order to estimate the amount of a certain type of material required to perform a job. The appropriate section of the technical specifications and page(s) of drawings should be carefully reviewed to determine the types and quantities of materials required. The quantities are then placed on the worksheet. For example, the specification section on finished carpentry should be reviewed along with the appropriate pages of drawings before taking off the linear feet of door and window trim.

If an estimate is required because not enough materials were ordered to complete the job, the estimator must also determine how much more work is necessary. Once this is known, the crew leader can then determine the materials needed. The construction drawings will also be used in this process.

WORKSHEET

Takeoff By:

Checked By:

DATE ____________

SHEET ______ of ______

PROJECT ____________________________

ARCHITECT ____________________________

PAGE #

REF.	DESCRIPTION	DIMENSIONS				EXTENSION	QUANTITY	UNIT	TOTAL		REMARKS
		NO.	LENGTH	WIDTH	HEIGHT				QUANTITY	UNIT	

46101-11_F12.EPS

Figure 12 Quantity takeoff sheet.

SUMMARY SHEET

By: _____

DATE _____ of _____

SHEET _____

TITLE: _____

PROJECT _____

WORK ORDER # _____

PAGE # _____

DESCRIPTION	QUANTITY		MATERIAL COST		LABOR MAN HOURS FACTORS					LABOR COST			ITEM COST	
	TOTAL	UT	PER UNIT	TOTAL	CRAFT	PR UNIT	TOTAL	RATE	COST PR	PER	TOTAL	TOTAL	PER UNIT
	MATERIAL									LABOR	TOTAL		

46101-11_F13.EPS

Figure 13 Summary sheet.

Assume you are the leader of a crew building footing formwork for the construction shown in *Figure 14*. You have used all of the materials provided for the job, yet you have not completed it. You study the drawings and see that the formwork consists of two side forms, each 12" high. The total length of footing for the entire project is 115'-0". You have completed 88'-0" to date; therefore, you have 27'-0" remaining (115'-0" – 88'-0" = 27'-0"). Your job is to prepare an estimate of materials that you will need to complete the job. In this case, only the side forms will be estimated (the miscellaneous materials will not be considered here).

* Footing length to complete: 27'-0"
* Footing height: 1'-0"

Refer to the worksheet in *Figure 15* for a final tabulation of the side forms needed to complete the job.

1. Using the same footing as described in the example above, calculate the quantity (square feet) of formwork needed to finish 203 linear feet of the footing. Place this information directly on the worksheet.
2. You are the crew leader of a carpentry crew whose task is to side a warehouse with plywood sheathing. The wall height is 16 feet, and there is a total of 480 linear feet of wall to side. You have done 360 linear feet of wall and have run out of materials. Calculate how many more feet of plywood you will need to complete the job. If you are using 4' × 8' plywood panels, how many will you need to order, assuming no waste? Write your estimate on the worksheet.

Show your calculations to the instructor.

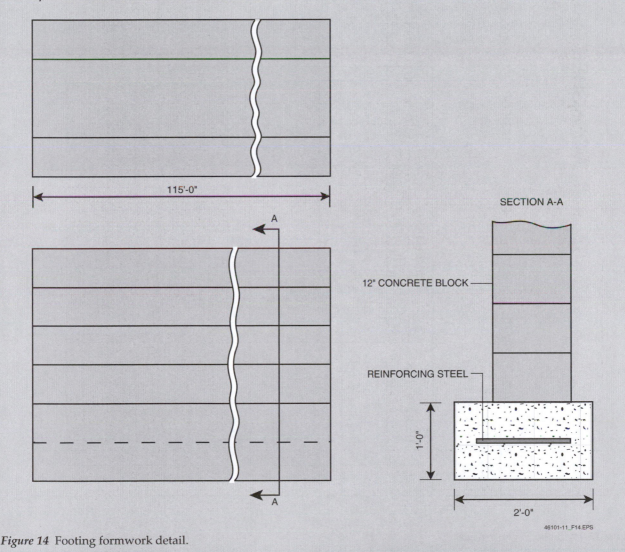

Figure 14 Footing formwork detail.

WORKSHEET

PAGE #1

Takeoff
By: RWH

DATE ___2/1/15___

Checked
By:

SHEET _01_ of _01_

PROJECT ___Sam's Diner___

ARCHITECT ___654b___

REF.	DESCRIPTION	NO	DIMENSIONS LENGTH	DIMENSIONS WIDTH	DIMENSIONS HEIGHT	EXTENSION	QUANTITY	UNIT	TOTAL QUANTITY	TOTAL UNIT	REMARKS
	Footing Side Forms 2		27'0"		1'0"	2x27x1	54	SF	54	SF	

46101-11_F15.EPS

Figure 15 Worksheet with entries.

4.0.0 PLANNING

Planning can be defined as determining the method used to carry out the different tasks to complete a project. It involves deciding what needs to be done and coming up with an organized sequence of events or plan for doing the work.

Planning involves the following:

- Determining the best method for performing the job
- Identifying the responsibilities of each person on the work crew
- Determining the duration and sequence of each activity
- Identifying what tools and equipment are needed to complete a job
- Ensuring that the required materials are at the work site when needed
- Making sure that heavy construction equipment is available when required

- Working with other contractors in such a way as to avoid interruptions and delays

4.1.0 Why Plan?

With a plan, a crew leader can direct work efforts efficiently and can use resources such as personnel, materials, tools, equipment, work area, and work methods to their full potential.

Some reasons for planning include the following:

- Controlling the job in a safe manner so that it is built on time and within cost
- Lowering job costs through improved productivity
- Preparing for bad weather or unexpected occurrences
- Promoting and maintaining favorable employee morale
- Determining the best and safest methods for performing the job

---

Participant Exercise F

1. In your own words, define planning, and describe how a job can be done better if it is planned. Give an example.

_____

_____

_____

_____

2. Consider a job that you recently worked on to answer the following:
   a. List the material(s) used.
   b. List each member of the crew with whom you worked and what each person did.
   c. List the kinds of equipment used.

_____

_____

_____

_____

_____

3. List some suggestions for how the job could have been done better, and describe how you would plan for each of the suggestions.

_____

_____

_____

_____

 Fundamentals of Crew Leadership

4.2.0 Stages of Planning

There are various times when planning is done for a construction job. The two most important occur in the pre-construction phase and during the construction work.

4.2.1 Pre-Construction Planning

The pre-construction stage of planning occurs before the start of construction. Except in a fairly small company or for a relatively small job, the crew leader usually does not get directly involved in the pre-construction planning process, but it is important to understand what it involves.

There are two phases of pre-construction planning. The first is when the proposal, bid, or negotiated price for the job is being developed. This is when the estimator, the project manager, and the field superintendent develop a preliminary plan for how the work will be done. This is accomplished by applying experience and knowledge from previous projects. It involves determining what methods, personnel, tools, and equipment will be used and what level of productivity they can achieve.

The second phase occurs after the contract is awarded. This phase requires a thorough knowledge of all project drawings and specifications. During this stage, the actual work methods and resources needed to perform the work are selected. Here, crew leaders might get involved, but their planning must adhere to work methods, production rates, and resources that fit within the estimate prepared before the contract was awarded. If the project requires a method of construction different from what is normal, the crew leader will usually be informed of what method to use.

4.2.2 Construction Planning

During construction, the crew leader is directly involved in planning on a daily basis. This planning consists of selecting methods for completing tasks before beginning work. Effective planning exposes likely difficulties, and enables the crew leader to minimize the unproductive use of personnel and equipment. Effective planning also provides a gauge to measure job progress. Effective crew leaders develop what is known as look-ahead (short-term) schedules. These schedules consider actual circumstances as well as projections two to three weeks into the future. Developing a look-ahead schedule helps ensure that all resources are available on the project when needed.

One of the characteristics of an effective crew leader is the ability to reduce each job to its simpler parts and organize a plan for handling each task.

Project planners establish time and cost limits for the project; the crew leader's planning must fit within those constraints. Therefore, it is important to consider the following factors that may affect the outcome:

- Site and local conditions, such as soil types, accessibility, or available staging areas
- Climate conditions that should be anticipated during the project
- Timing of all phases of work
- Types of materials to be installed and their availability
- Equipment and tools required and their availability
- Personnel requirements and availability
- Relationships with the other contractors and their representatives on the job

On a simple job, these items can be handled almost automatically. However, larger or more complex jobs require the planner to give these factors more formal consideration and study.

5.0.0 THE PLANNING PROCESS

The planning process consists of the following five steps:

Step 1 Establish a goal.

Step 2 Identify the work activities that must be completed in order to achieve the goal.

Step 3 Identify the tasks that must be done to accomplish those activities.

Step 4 Communicate responsibilities.

Step 5 Follow up to see that the goal is achieved.

5.1.0 Establish a Goal

The term *goal* has different meanings for different people. In general, a goal is a specific outcome that one works toward. For example, the project superintendent of a home construction project could establish the goal to have a house dried-in by a certain date. (Dried-in means ready for the application of roofing and siding.) In order to meet that goal, the leader of the framing crew and the superintendent would need to agree to a goal to have the framing completed by a given date. The crew leader would then establish sub-goals (objectives) for the crew to complete each element of the framing (floors, walls, roof) by a set time. The superintendent would need to set similar goals with the crews that install sheathing, building wrap, windows, and exterior doors. However, if the framing crew does not meet its goal, the other crews will be delayed.

5.2.0 Identify the Work to be Done

The second step in planning is to identify the work to be done to achieve the goal. In other words, it is a series of activities that must be done in a certain sequence. The topic of breaking down a job into activities is covered later in this section. At this point, the crew leader should know that, for each activity, one or more objectives must be set.

An objective is a statement of what is desired at a specific time. An objective must:

- Mean the same thing to everyone involved
- Be measurable, so that everyone knows when it has been reached
- Be achievable
- Have everyone's full support

Examples of practical objectives include the following:

- By 4:30 P.M. today, the crew will have completed installation of the floor joists.
- By closing time Friday, the roof framing will be complete.

Notice that both examples meet the first three requirements of an objective. In addition, it is assumed that everyone involved in completing the task is committed to achieving the objective. The advantage in developing objectives for each work activity is that it allows the crew leader to evaluate whether or not the plan and schedules are being followed. In addition, objectives serve as sub-goals that are usually under the crew leader's control.

Some construction work activities, such as installing 12"-deep footing forms, are done so often that they require little planning. However, other jobs, such as placing a new type of mechanical equipment, require substantial planning. This type of job requires that the crew leader set specific objectives.

Whenever faced with a new or complex activity, take the time to establish objectives that will serve as guides for accomplishing the job. These guides can be used in the current situation, as well as in similar situations in the future.

5.3.0 Identify Tasks to be Performed

To plan effectively, the crew leader must be able to break a work activity assignment down into smaller tasks. Large jobs include a greater number of tasks than small ones, but all jobs can be broken down into manageable components.

When breaking down an assignment into tasks, make each task identifiable and definable. A task is identifiable when the types and amounts of resources it requires are known. A task is definable if it has a specific duration. For purposes of efficiency, the job breakdown should not be too detailed or complex, unless the job has never been done before or must be performed with strictest efficiency.

For example, a suitable breakdown for the work activity to install 12" × 12" vinyl floor tile in a cafeteria might be the following:

Step 1   Prepare the floor.

Step 2   Lay out the tile.

Step 3   Spread the adhesive.

Step 4   Lay the tile.

Step 5   Clean the tile.

Step 6   Wax the floor.

The crew leader could create even more detail by breaking down any one of the tasks, such as lay the tile, into subtasks. In this case, however, that much detail is unnecessary and wastes the crew leader's time and the project's money. However, breaking tasks down further might be necessary in a case where the job is very complex or the analysis of the job needs to be very detailed.

Every work activity can be divided into three general parts:

- Preparing
- Performing
- Cleaning up

One of the most frequent mistakes made in the planning process is forgetting to prepare and to clean up. The crew leader must be certain that preparation and cleanup are not overlooked.

After identifying the various tasks that make up the job and developing an objective for each task, the crew leader must determine what resources the job requires. Resources include labor, equipment, materials, and tools. In most jobs, these resources are identified in the job estimate. The crew leader must make sure that these resources are available on the site when needed.

5.4.0 Communicating Responsibilities

A crew leader is unable to complete all of the activities within a job independently. Other people must be relied upon to get everything done. Therefore, most jobs have a crew of people with various experiences and skill levels to assist in the work. The crew leader's job is to draw from this expertise to get the job done well and in a safe and timely manner.

Once the various activities that make up the job have been determined, the crew leader must identify the person or persons responsible for completing each activity. This requires that the crew leader be aware of the skills and abilities of the people on the crew. Then, the crew leader must put this knowledge to work in matching the crew's skills and abilities to specific tasks that must be performed to complete the job.

After matching crew members to specific activities, the crew leader must then communicate the assignments to the crew. Communication of responsibilities is generally handled verbally; the crew leader often talks directly to the person to which the activity has been assigned. There may be times when work is assigned indirectly through written instructions or verbally through someone other than the crew leader. Either way, the crew members should know what it is they are responsible for accomplishing on the job.

5.5.0 Follow-Up Activities

Once the activities have been delegated to the appropriate crew members, the crew leader must follow up to make sure that they are completed effectively and efficiently. Follow-up work involves being present on the job site to make sure all the resources are available to complete the work; ensuring that the crew members are working on their assigned activities; answering any questions; and helping to resolve any problems that occur while the work is being done. In short, follow-up activity means that the crew leader is aware of what's going on at the job site and is doing whatever is necessary to make sure that the work is completed on schedule.

Figure 16 reviews the planning steps.

The crew leader should carry a small note pad or electronic device to be used for planning and note taking. That way, thoughts about the project can be recorded as they occur, and pertinent details will not be forgotten. The crew leader may also choose to use a planning form such as the one illustrated in *Figure 17*.

As the job is being built, refer back to these plans and notes to see that the tasks are being done in sequence and according to plan. This is referred to as analyzing the job. Experience shows that jobs that are not built according to work plans usually end up costing more and taking more time; therefore, it is important that crew leaders refer back to the plans periodically.

The crew leader is involved with many activities on a day-to-day basis. As a result, it is easy to forget important events if they are not documented. To help keep track of events such as job changes, interruptions, and visits, the crew leader should keep a job diary.

Figure 16 Steps to effective planning.

DAILY WORK PLAN

"PLAN YOUR WORK AND WORK YOUR PLAN = EFFICIENCY"

Plan of _____   Date _____

PRIORITY	DESCRIPTION	✓ When Completed ✗ Carried Forward

46101-11_F17.EPS

Figure 17 Planning form.

A job diary is a notebook in which the crew leader records activities or events that take place on the job site that may be important later. When recording in a job diary, make sure that the information is accurate, factual, complete, consistent, organized, and up to date. Follow company policy in determining which events should be recorded. However, if there is a doubt about what to include, it is better to have too much information than too little.

Figure 18 shows a sample page from a job diary.

6.0.0 PLANNING RESOURCES

Once a job has been broken down into its tasks or activities, the next step is to assign the various resources needed to perform them.

6.1.0 Safety Planning

Using the company safety manual as a guide, the crew leader must assess the safety issues associated with the job and take necessary measures to minimize any risk to the crew. This may involve working with the company or site safety officer and may require a formal hazard analysis.

6.2.0 Materials Planning

The materials required for the job are identified during pre-construction planning and are listed on the job estimate. The materials are usually ordered from suppliers who have previously provided quality materials on schedule and within estimated cost.

July 8, 2015

Weather: Hot and Humid

Project: Company XYZ Building

- The paving contractor crew arrived late (10 am).

- The owner representative inspected the footing foundation at approximately 1 pm.

- The concrete slump test did not pass. Two trucks had to be ordered to return to the plant, causing a delay.

- John Smith had an accident on the second floor. I sent him to the doctor for medical treatment. The cause of the accident is being investigated.

46101-11_F18.EPS

Figure 18 Sample page from a job diary.

The crew leader is usually not involved in the planning and selection of materials, since this is done in the pre-construction phase. The crew leader does, however, have a major role to play in the receipt, storage, and control of the materials after they reach the job site.

The crew leader is also involved in planning materials for tasks such as job-built formwork and scaffolding. In addition, the crew leader may run out of a specific material, such as fasteners, and need to order more. In such cases, a higher authority should be consulted, since most companies have specific purchasing policies and procedures.

6.3.0 Site Planning

There are many planning elements involved in site work. The following are some of the key elements:

- Emergency procedures
- Access roads
- Parking
- Stormwater runoff
- Sedimentation control
- Material and equipment storage
- Material staging
- Site security

6.4.0 Equipment Planning

Much of the planning for use of construction equipment is done during the pre-construction phase. This planning includes the types of equipment needed, the use of the equipment, and the length of time it will be on the site. The crew leader must work with the home office to make certain that the equipment reaches the job site on time. The crew leader must also ensure that sure equipment operators are properly trained.

Coordinating the use of the equipment is also very important. Some equipment is used in combination with other equipment. For example, dump trucks are generally required when loaders and excavators are used. The crew leader should also coordinate equipment with other contractors on the job. Sharing equipment can save time and money and avoid duplication of effort.

46101-11_SA03.EPS

The crew leader must designate time for equipment maintenance in order to prevent equipment failure. In the event of an equipment failure, the crew leader must know who to contact to resolve the problem. An alternate plan must be ready in case one piece of equipment breaks down, so that the other equipment does not sit idle. This planning should be done in conjunction with the home office or the crew leader's immediate superior.

6.5.0 Tool Planning

A crew leader is responsible for planning what tools will be used on the job. This task includes:

- Determining the tools required
- Informing the workers who will provide the tools (company or worker)
- Making sure the workers are qualified to use the tools safely and effectively
- Determining what controls to establish for tools

6.6.0 Labor Planning

All jobs require some sort of labor because the crew leader cannot complete all the work alone. When planning for labor, the crew leader must:

- Identify the skills needed to perform the work.
- Determine how many people having those specific skills are needed.
- Decide who will actually be on the crew.

In many companies, the project manager or job superintendent determines the size and makeup of the crew. Then, the crew leader is expected to accomplish the goals and objectives with the crew provided. Even though the crew leader may not be involved in staffing the crew, the crew leader is responsible for training the crew members to ensure that they have the skills needed to do the job.

In addition, the crew leader is responsible for keeping the crew adequately staffed at all times so that jobs are not delayed. This involves dealing with absenteeism and turnover, two common problems that affect industry today.

7.0.0 SCHEDULING

Planning and scheduling are closely related and are both very important to a successful job. Planning involves determining the activities that must be completed and how they should be accomplished. Scheduling involves establishing start and finish times or dates for each activity.

A schedule for a project typically shows:

- Operations listed in sequential order
- Units of construction
- Duration of activities
- Estimated date to start and complete each activity
- Quantity of materials to be installed

There are different types of schedules used today. They include the bar chart; the network schedule, which is sometimes called the critical path method (CPM) or precedence diagram; and the short-term, or look-ahead schedule.

7.1.0 The Scheduling Process

The following is a brief summary of the steps a crewleader must complete to develop a schedule.

Step 1 Make a list of all of the activities that will be performed to build the job, including individual work activities and special tasks, such as inspections or the delivery of materials.

At this point, the crew leader should just be concerned with generating a list, not with determining how the activities will be accomplished, who will perform them, how long they will take, or in what sequence they will be completed.

Step 2 Use the list of activities created in Step 1 to reorganize the work activities into a logical sequence.

When doing this, keep in mind that certain steps cannot happen until others have been completed. For example, footings must be excavated before concrete can be placed.

Step 3 Assign a duration or length of time that it will take to complete each activity and determine the start time for each. Each activity will then be placed into a schedule format. This step is important because it helps the crew leader ensure that the activities are being completed on schedule.

The crew leader must be able to read and interpret the job schedule. On some jobs, the beginning and expected end date for each activity, along with the expected crew or worker's production rate, is provided on the form. The crew leader can use this

information to plan work more effectively, set realistic goals, and measure whether or not they were accomplished within the scheduled time.

Before starting a job, the crew leader must:

- Determine the materials, tools, equipment, and labor needed to complete the job.
- Determine when the various resources are needed.
- Follow up to ensure that the resources are available on the job site when needed.

Availability of needed resources should be verified three to four working days before the start of the job. It should be done even earlier for larger jobs. This advance preparation will help avoid situations that could potentially delay the job or cause it to fall behind schedule.

7.2.0 Bar Chart Schedule

Bar chart schedules, also known as Gantt charts, can be used for both short-term and long-term jobs. However, they are especially helpful for jobs of short duration.

Bar charts provide management with the following:

- A visual concept of the overall time required to complete the job through the use of a logical method rather than a calculated guess
- A means to review each part of the job
- Coordination requirements between crafts
- Alternative methods of performing the work

A bar chart can be used as a control device to see whether the job is on schedule. If the job is not on schedule, immediate action can be taken in the office and the field to correct the problem and ensure that the activity is completed on schedule.

A bar chart is illustrated in *Figure 19*.

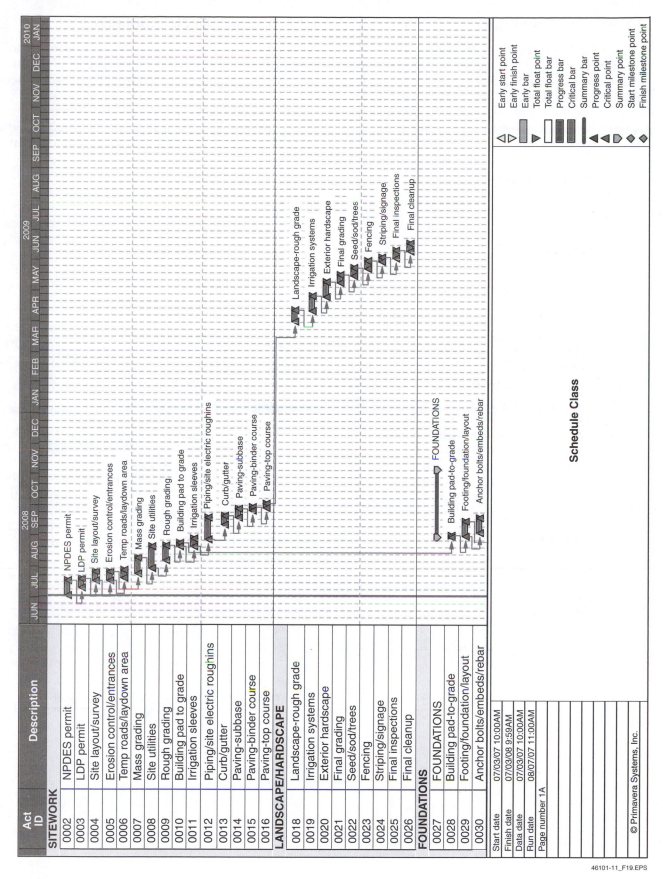

Figure 19 Example of a bar chart schedule.

46101-11_F19.EPS

7.3.0 Network Schedule

Network schedules are an effective project management tool because they show dependent (critical path) activities and activities that can be performed in parallel. In *Figure 20*, for example, reinforcing steel cannot be set until the concrete forms have been built and placed. Other activities are happening in parallel, but the forms are in the critical path. When building a house, drywall cannot be installed and finished until wiring, plumbing, and HVAC ductwork have been roughed-in. Because other activities, such as painting and trim work, depend on drywall completion, the drywall work is a critical-path function. That is, until it is complete, the other tasks cannot be started, and the project itself is likely to be delayed by the amount of delay in any dependent activity. Likewise, drywall work can't even be started until the rough-ins are complete. Therefore, the project superintendent is likely to focus on those activities when evaluating schedule performance.

The advantage of a network schedule is that it allows project leaders to see how a schedule change on one activity is likely to affect other activities and the project in general. A network schedule is laid out on a timeline and usually shows the estimated duration for each activity. Network schedules are generally used for complex jobs that take a long time to complete. The PERT (program evaluation and review technique) schedule is a form of network schedule.

7.4.0 Short-Term Scheduling

Since the crew leader needs to maintain the job schedule, he or she needs to be able to plan daily production. Short-term scheduling is a method used to do this. An example is shown in *Figure 21*.

The information to support short-term scheduling comes from the estimate or cost breakdown. The schedule helps to translate estimate data and the various job plans into a day-to-day schedule of events. The short-term schedule provides the crew leader with visibility over the project. If actual production begins to slip behind estimated production, the schedule will warn the crew leader that a problem lies ahead and that a schedule slippage is developing.

Short-term scheduling can be used to set production goals. It is generally agreed that production can be improved if workers:

- Know the amount of work to be accomplished
- Know the time they have to complete the work
- Can provide input when setting goals

Consider the following example:

Situation:

A carpentry crew on a retaining wall project is about to form and pour catch basins and put up wall forms. The crew has put in a number of catch basins, so the crew leader is sure that they can perform the work within the estimate. However, the crew leader is concerned about their production of the wall forms. The crew will work on both the basins and the wall forms at the same time.

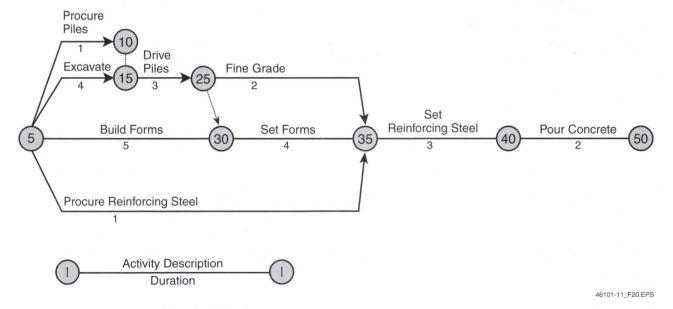

Figure 20 Example of a network schedule.

	Calendar Dates															
	7/1	7/2	7/3	7/7	7/8	7/9	7/10	7/11	7/14	7/15	7/16	7/17	7/18	7/21	7/22	7/23
ACTIVITY DESCRIPTION — Work Days	1	2	3	4	5	6	7	8	9	10	11	12	13	14	15	16
Process Piles	▓															
Excavate	▓	▓	▓	▓												
Build Forms	▓	▓	▓	▓	▓											
Process Reinforcing Steel	▓															
Drive Piles					▓	▓	▓									
Fine Grade								▓	▓							
Set Forms								▓	▓	▓	▓					
Set Reinforced Steel												▓	▓	▓		
Pour Concrete															▓	▓

NOTES: The project start date is July 1st, which is a Tuesday.
Time placement of activity and duration may be done anytime through shaded portion.
Bottom portion of line available to show progress as activities are completed.

ADDITIONAL TASKS:

					▓											

Figure 21 Short-term schedule.

1. The crew leader notices the following in the estimate or cost breakdown:
   a. Production factor for wall forms = 16 work-hours per 100 square feet
   b. Work to be done by measurement = 800 square feet
   c. Total time = 128 work-hours (800 × 16 ÷ 100)
2. The carpenter crew consists of the following:
   a. One carpenter crew leader
   b. Four carpenters
   c. One laborer
3. The crew leader determines the goal for the job should be set at 128 work-hours (from the cost breakdown).
4. If the crew remains the same (six workers), the work should be completed in about 21 crew-hours (128 work hours ÷ 6 workers = 21.33 crew-hours).
5. The crew leader then discusses the production goal (completing 800 square feet in 21 crew-hours) with the crew and encourages them to work together to meet the goal of getting the forms erected within the estimated time.

The short-term schedule was used to translate production into work-hours or crew-hours and to schedule work so that the crew can accomplish it within the estimate. In addition, setting production targets provides the motivation to produce more than the estimate requires.

7.5.0 Updating a Schedule

No matter what type of schedule is used, it must be kept up to date to be useful to the crew leader. Inaccurate schedules are of no value.

The person responsible for scheduling in the office handles the updates. This person uses information gathered from job field reports to do the updates.

The crew leader is usually not directly involved in updating schedules. However, he or she may be responsible for completing field or progress reports used by the company. It is critical that the crew leader fill out any required forms or reports completely and accurately so that the schedule can be updated with the correct information.

8.0.0 COST CONTROL

Being aware of costs and controlling them is the responsibility of every employee on the job. It is the crew leader's job to ensure that employees uphold this responsibility. Control refers to the comparison of estimated performance against actual performance and following up with any needed corrective action. Crew leaders who use cost-control practices are more valuable to the company than those who do not.

On a typical job, many activities are going on at the same time. This can make it difficult to control the activities involved. The crew leader must be constantly aware of the costs of a project and effectively control the various resources used on the job.

When resources are not controlled, the cost of the job increases. For example, a plumbing crew of four people is installing soil pipe and does not have enough fittings. Three crew members wait while one crew member goes to the supply house for a part that costs only a few dollars. It takes the crew member an hour to get the part, so four hours of productive work have been lost. In addition, the travel cost for retrieving the supplies must be added.

8.1.0 Assessing Cost Performance

Cost performance on a project is determined by comparing actual costs to estimated costs. Regardless of whether the job is a contract bid project or an in-house project, a budget must first be established. In the case of a contract bid, the budget is generally the cost estimate used to bid the job. For an in-house job, participants will submit labor and material forecasts, and someone in authority will authorize a project budget.

It is common to estimate cost by either breaking the job into funded tasks or by forecasting labor and materials expenditures on a timeline. Many companies create a work breakdown structure (WBS) for each project. Within the WBS, each major task is assigned a discrete charge number. Anyone working on that task charges that number on their time sheet, so that project managers can readily track cost performance. However, knowing how much has been spent does not necessarily determine cost performance.

Although financial reports can show that actual expenses are tracking forecast expenses, they don't show if the work is being done at the required rate. Thus it is possible to have spent half the budget, but have less than half of the work compete. When the project is broken down into funded tasks related to schedule activities and events, there is far greater control over cost performance.

8.2.0 Field Reporting System

The total estimated cost comes from the job estimates, but the actual cost of doing the work is obtained from an effective field reporting system.

A field reporting system is made up of a series of forms, which are completed by the crew leader and others. Each company has its own forms and methods for obtaining information. The general information and the process of how they are used are described here.

First, the number of hours each person worked on each task must be known. This information comes from daily time cards. Once the accounting department knows how many hours each employee worked on an activity, it can calculate the total cost of the labor by multiplying the number of hours worked by the wage rate for each worker. The cost for the labor to do each task can be calculated as the job progresses. This cost will be compared with the estimated cost. This comparison will also be done at the end of the job.

When material is put in place, a designated person will measure the quantities from time to time and send this information to the home office and, possibly, the crew leader. This information, along with the actual cost of the material and the amount of hours it took the workers to install it, is compared to the estimated cost. If the cost is greater than the estimate, management and the crew leader will have to take action to reduce the cost.

A similar process is used to determine if the costs to operate equipment or the production rate are comparable to the estimated cost and production rate.

For this comparison process to be of use, the information obtained from field personnel must be correct. It is important that the crew leader be accurate in reporting. The crew leader is responsible for carrying out his or her role in the field reporting system. One of the best ways to do this is to maintain a daily diary, using a notebook or electronic device. In the event of a legal/contractual conflict with the client, such diaries are considered as evidence in court proceedings, and can be helpful in reaching a settlement.

Here is an example. You are running a crew of five concrete finishers for a subcontractor. When you and your crew show up to finish a slab, the GC says, "We're a day behind on setting the forms, so I need you and your crew to stand down until tomorrow." What do you do?

Of course, you would first call your office to let them know about the delay. Then, you would immediately record it in your job diary. A six-man crew for one day represents 48 labor hours. If your company charges $30 an hour, that's a potential loss of $1,440, which the company would want to recover from the GC. If there is a dispute, your entry in the job diary could result in a favorable decision for your employer.

8.3.0 Crew Leader's Role in Cost Control

The crew leader is often the company representative in the field, where the work takes place. Therefore, the crew leader has a great deal to do with determining job costs. When work is assigned to a crew, the crew leader should be given a budget and schedule for completing the job. It is then up to the crew leader to make sure the job is done on time and stays within budget. This is done by actively managing the use of labor, materials, tools, and equipment.

If the actual costs are at or below the estimated costs, the job is progressing as planned and scheduled, and the company will realize the expected profit. However, if the actual costs exceed the estimated costs, one or more problems may result in the company losing its expected profit, and maybe more. No company can remain in business if it continually loses money. One of the factors that can increase cost is client-related changes. The crew leaders must be able to assess the potential impact of such changes and, if necessary, confer with their employer to determine the course of action. If losses are occurring, the crew leader and superintendent will need to work together to get the costs back in line.

Noted below are some of the reasons why actual costs can exceed estimated costs and suggestions on what the crew leader can do to bring the costs back in line. Before starting any action, however, the crew leader should check with his or her superior to see that the action proposed is acceptable and within the company's policies and procedures.

- *Cause* – Late delivery of materials, tools, and/or equipment
  Corrective Action: Plan ahead to ensure that job resources will be available when needed
- *Cause* – Inclement weather
  Corrective Action: Work with the superintendent and have alternate plans ready
- *Cause* – Unmotivated workers
  Corrective Action: Counsel the workers
- *Cause* – Accidents
  Corrective Action: Enforce the existing safety program

There are many other methods to get the job done on time if it gets off schedule. Examples include working overtime, increasing the size of the crew, pre-fabricating assemblies, or working staggered shifts. However, these examples may increase the cost of the job, so they should not be done without the approval of the project manager.

9.0.0 RESOURCE CONTROL

The crew leader's job is to ensure that assigned tasks are completed safely according to the plans and specifications, on schedule, and within the scope of the estimate. To accomplish this, the crew leader must closely control how resources of materials, equipment, tools, and labor are used. Waste must be minimized whenever possible.

Control involves measuring performance and correcting deviations from plans and specifications to accomplish objectives. Control anticipates deviation from plans and specifications and takes measures to prevent it from occurring.

An effective control process can be broken down into the following steps:

Step 1   Establish standards and divide them into measurable units.

For example, a baseline can be created using experience gained on a typical job, where 2,000 LF of 1¼" copper water tube was installed in five days. Dividing 2,000 by 5 gives 400. Thus, the average installation rate in this case for 1¼" copper water tube was 400 LF/day.

Step 2   Measure performance against a standard.

On another job, 300 square feet of the same tube was placed during an average day. Thus, this average production of 300 LF/day did not meet the average rate of 400 LF/day.

Step 3   Adjust operations to ensure that the standard is met.

In Step 2 above, if the plan called for the job to be completed in five days, the crew leader would have to take action to ensure that this happens. If 300 LF/day is the actual average daily rate, it will have to be increased by 100 LF/day to meet the standard.

9.1.0 Materials Control

The crew leader's responsibility in materials control depends on the policies and procedures of the company. In general, the crew leader is responsible for ensuring on-time delivery, preventing waste, controlling delivery and storage, and preventing theft of materials.

9.1.1 Ensuring On-Time Delivery

It is essential that the materials required for each day's work be on the job site when needed. The crew leader should confirm in advance that all materials have been ordered and will be delivered on schedule. A week or so before the delivery date, follow-up is needed to make sure there will be no delayed deliveries or items on backorder.

To be effective in managing materials, the crew leader must be familiar with the plans and specifications to be used, as well as the activities to be performed. He or she can then determine how many and what types of materials are needed.

If other people are responsible for providing the materials for a job, the crew leader must follow up to make sure that the materials are available when they are needed. Otherwise, delays occur as crew members stand around waiting for the materials to be delivered.

9.1.2 Preventing Waste

Waste in construction can add up to loss of critical and costly materials and may result in job delays. The crew leader needs to ensure that every crew member knows how to use the materials efficiently. The crew should be monitored to make certain that no materials are wasted.

An example of waste is a carpenter who saws off a piece of lumber from a fresh piece, when the size needed could have been found in the scrap pile. Another example of waste involves installing a fixture or copper tube incorrectly. The time spent installing the item incorrectly is wasted because the task will need to be redone. In addition, the materials may need to be replaced if damaged during installation or removal.

Under LEED, waste control is very important. Credits are given for finding ways to reduce waste and for recycling waste products. Waste material should be separated for recycling if feasible (*Figure 22*).

Did you know?

Just-in-time (JIT) delivery is a strategy in which materials are delivered to the job site when they are needed. This means that the materials may be installed right off the truck. This method reduces the need for on-site storage and staging. It also reduces the risk of loss or damage as products are moved about the site. Other modern material management methods include the use of radio frequency identification tags (RFIDs) that make it easy to locate material in crowded staging areas.

9.1.3 Verifying Material Delivery

A crew leader may be responsible for the receipt of materials delivered to the work site. When this happens, the crew leader should require a copy of the shipping ticket and check each item on the shipping ticket against the actual materials to see that the correct amounts were received.

Figure 22 Waste material separated for recycling.

The crew leader should also check the condition of the materials to verify that nothing is defective before signing the shipping ticket. This can be difficult and time consuming because it means that cartons must be opened and their contents examined. However, it is necessary, because a signed shipping ticket indicates that all of the materials were received undamaged. If the crew leader signs for the materials without checking them, and then finds damage, no one will be able to prove that the materials came to the site in that condition.

Once the shipping ticket is checked and signed, the crew leader should give the original or a copy to the superintendent or project manager. The shipping ticket will then be filed for future reference because it serves as the only record the company has to check bills received from the supply house.

9.1.4 Controlling Delivery and Storage

Another very important element of materials control is where the materials will be stored on the job site. There are two factors in determining the appropriate storage location. The first is convenience. If possible, the materials should be stored near where they will be used. The time and effort saved by not having to carry the materials long distances will greatly reduce the installation costs.

Next, the materials must be stored in a secure area where they will not be damaged. It is important that the storage area suit the materials being stored. For instance, materials that are sensitive to temperature, such as chemicals or paints, should be stored in climate-controlled areas. Otherwise, waste may occur.

9.1.5 Preventing Theft and Vandalism

Theft and vandalism of construction materials increase costs because these materials are needed to complete the job. The replacement of materials and the time lost because the needed materials are missing can add significantly to the cost. In addition, the insurance that the contractor purchases will increase in cost as the theft and vandalism rate grows.

The best way to avoid theft and vandalism is a secure job site. At the end of each work day, store unused materials and tools in a secure location, such as a locked construction trailer. If the job site is fenced or the building can be locked, the materials can be stored within. Many sites have security cameras and/or intrusion alarms to help minimize theft and vandalism.

9.2.0 Equipment Control

The crew leader may not be responsible for long-term equipment control. However, the equipment required for a specific job is often the crew leader's responsibility. The first step is to identify when the required equipment must be transported from the shop or rental yard. The crew leader is responsible for informing the shop where it is being used and seeing that it is returned to the shop when the job is done.

It is common for equipment to lay idle at a job site because the job has not been properly planned and the equipment arrived early. For example, if wire-pulling equipment arrives at a job site before the conduit is in place, this equipment will be out of service while awaiting the conduit installation. In addition, it is possible that this unused equipment could be damaged, lost, or stolen.

The crew leader needs to control equipment use, ensure that the equipment is operated in accordance with its design, and that it is being used within time and cost guidelines. The crew leader must also ensure that equipment is maintained and repaired as indicated by the preventive maintenance schedule. Delaying maintenance and repairs can lead to costly equipment failures. The crew leader must also ensure that the equipment operators have the necessary credentials to operate the equipment, including applicable licenses.

The crew leader is responsible for the proper operation of all other equipment resources, including cars and trucks. Reckless or unsafe operation of vehicles will likely result in damaged equipment and a delayed or unproductive job. This, in turn, could affect the crew leader's job security.

The crew leader should also ensure that all equipment is secured at the close of each day's work in an effort to prevent theft. If the equipment is still being used for the job, the crew leader should ensure that it is locked in a safe place; otherwise, it should be returned to the shop.

9.3.0 Tool Control

Among companies, various policies govern who provides hand and power tools to employees. Some companies provide all the tools, while others furnish only the larger power tools. The crew leader should find out about and enforce any company policies related to tools.

Tool control is a twofold process. First, the crew leader must control the issue, use, and maintenance of all tools provided by the company. Next, the crew leader must control how the tools are being used to do the job. This applies to tools that are issued by the company as well as tools that belong to the workers.

Using the proper tools correctly saves time and energy. In addition, proper tool use reduces the chance of damage to the tool being used. Proper use also reduces injury to the worker using the tool, and to nearby workers.

Tools must be adequately maintained and properly stored. Making sure that tools are cleaned, dried, and lubricated prevents rust and ensures that the tools are in the proper working order.

In the event that tools are damaged, it is essential that they be repaired or replaced promptly. Otherwise, an accident or injury could result.

> **NOTE**
>
> Regardless of whether a tool is owned by a worker or the company, OSHA will hold the company responsible for it when it is used on a job site. The company can be held accountable if an employee is injured by a defective tool. Therefore, the crew leader needs to be aware of any defects in the tools the crew members are using.

Company-issued tools should be taken care of as if they are the property of the user. Workers should not abuse tools simply because they are not their own.

One of the major causes of time lost on a job is the time spent searching for a tool. To prevent this from occurring, a storage location for company-issued tools and equipment should be established. The crew leader should make sure that all company-issued tools and equipment are returned to this designated location after use. Similarly, workers should make sure that their personal toolboxes are organized so that they can readily find the appropriate tools and return their tools to their toolboxes when they are finished using them.

Studies have shown that the key to an effective tool control system lies in:

- Limiting the number of people allowed access to stored tools
- Limiting the number of people held responsible for tools
- Controlling the ways in which a tool can be returned to storage
- Making sure tools are available when needed

9.4.0 Labor Control

Labor typically represents more than half the cost of a project, and therefore has an enormous impact on profitability. For that reason, it is essential to manage a crew and their work environment in a way that maximizes their productivity. One of the ways to do that is to minimize the time spent on unproductive activities such as:

- Engaging in bull sessions
- Correcting drawing errors
- Retrieving tools, equipment, and materials
- Waiting for other workers to finish

If crew members are habitually goofing off, it is up to the crew leader to counsel those workers. The counseling should be documented in the crew leader's daily diary. Repeated violations will need to be referred to the attention of higher management as guided by company policy.

Errors will occur and will need to be corrected. Some errors, such as mistakes on drawings, may be outside of the crew leader's control. However, some drawing errors can be detected by carefully examining the drawings before work begins.

If crew members are making mistakes due to inexperience, the crew leader can help avoid these errors by providing on-the-spot training and by checking on inexperienced workers more often.

The availability and location of tools, equipment, and materials can have a profound effect on a crew's productivity. If the crew has to wait for these things, or travel a distance to get them, it reduces their productivity. The key to minimizing such problems is proactive management of these resources. As discussed earlier, practices such as checking in advance to be sure equipment and materials will be available when scheduled and placing materials close to the work site will help minimize unproductive time.

Delays caused by others can be avoided by carefully tracking the project schedule. By doing so, crew leaders can anticipate delays that will affect the work of their crews and either take action to prevent the delay or redirect the crew to another task.

Participant Exercise G

1. List the methods your company uses to minimize waste.

_____

_____

_____

_____

2. List the methods your company uses to control small tools on the job.

_____

_____

_____

_____

3. List five ways that you feel your company could control labor to maximize productivity.

_____

_____

_____

_____

_____

10.0.0 PRODUCTION AND PRODUCTIVITY

Production is the amount of construction put in place. It is the quantity of materials installed on a job, such as 1,000 linear feet of waste pipe installed in a given day. On the other hand, productivity depends on the level of efficiency of the work. It is the amount of work done per hour or day by one worker or a crew.

Production levels are set during the estimating stage. The estimator determines the total amount of materials to be put in place from the plans and specifications. After the job is complete, the actual amount of materials installed can be assessed, and the actual production can be compared to the estimated production.

Productivity relates to the amount of materials put in place by the crew over a certain time period. The estimator uses company records during the estimating stage to determine how much time and labor it will take to place a certain quantity of materials. From this information, the estimator calculates the productivity necessary to complete the job on time.

For example, it might take a crew of two people ten days to paint 5,000 square feet. The crew's productivity per day is obtained by dividing 5,000 square feet by ten days. The result is 500 square feet per day. The crew leader can compare the daily production of any crew of two painters doing similar work with this average, as discussed previously.

Planning is essential to productivity. The crew must be available to perform the work, and have all of the required materials, tools, and equipment in place when the job begins.

The time on the job should be for business, not for taking care of personal problems. Anything not work-related should be handled after hours, away from the job site. Planning after-work activities, arranging social functions, or running personal errands should be handled after work or during breaks.

Organizing field work can save time. The key to effectively using time is to work smarter, not necessarily harder. For example, most construction projects require that the contractor submit a set of as-built plans at the completion of the work. These plans describe how the materials were actually installed. The best way to prepare these plans is to mark a set of working plans as the work is in progress. That way, pertinent details will not be forgotten and time will not be wasted trying to remember how the work was done.

The amount of material actually used should not exceed the estimated amount. If it does, either the estimator has made a mistake, undocumented changes have occurred, or rework has caused the need for additional materials. Whatever the case, the crew leader should use effective control techniques to ensure the efficient use of materials.

When bidding a job, most companies calculate the cost per labor hour. For example, a ten-day job might convert to 160 labor hours (two painters for ten days at eight hours per day). If the company charges a labor rate of $30/hour, the labor cost would be $4,800. The estimator then adds the cost of materials, equipment, and tools, along with overhead costs and a profit factor, to determine the price of the job.

After a job has been completed, information gathered through field reporting allows the home office to calculate actual productivity and compare it to the estimated figures. This helps to identify productivity issues and improves the accuracy of future estimates.

The following labor-related practices can help to ensure productivity:

- Ensure that all workers have the required resources when needed.
- Ensure that all personnel know where to go and what to do after each task is completed.
- Make reassignments as needed.
- Ensure that all workers have completed their work properly.

1. Which of these activities occurs during the development phase of a project?
   a. Architect/engineer sketches are prepared and a preliminary budget is developed.
   b. Government agencies give a final inspection of the design, adherence to codes, and materials used.
   c. Project drawings and specifications are prepared.
   d. Contracts for the project are awarded.

2. The type of contract in which the client pays the contractor for their actual labor and material expenses they incur is known as a _____.
   a. firm fixed-price contract
   b. time-spent contract
   c. cost-reimbursable contract
   d. performance-based contract

3. On-site changes in the original design that are made during construction are recorded in the _____.
   a. as-built plans
   b. takeoff sheet
   c. project schedule
   d. job specifications

4. On a design-build project, _____.
   a. the owner is responsible for providing the design
   b. the architect does the design and the general contractor builds the project
   c. the same contractor is responsible for both design and construction
   d. a construction manager is hired to oversee the project

5. One example of a direct cost when bidding a job is _____.
   a. office rent
   b. labor
   c. accounting
   d. utilities

6. The control method that a crew leader uses to plan a few weeks in advance is a _____.
   a. network schedule
   b. bar chart schedule
   c. daily diary
   d. look-ahead schedule

7. A job diary should typically indicate _____.
   a. items such as job interruptions and visits
   b. changes needed to project drawings
   c. the estimated time for each job task related to a particular project
   d. the crew leader's ideas for improving employee morale

8. Gantt charts can help crew leaders in the field by _____.
   a. offering a comparison of actual production to estimated production
   b. providing short-term and long-term schedule information
   c. stating the equipment and materials necessary to complete a task
   d. providing the information needed to develop an estimate or an estimate breakdown

9. What is the crew leader's responsibility with regard to cost control?
   a. Cost control is outside the scope of a crew leader's responsibility.
   b. The crew leader is responsible only for minimizing material waste.
   c. The crew leader must ensure that all team members are aware of project costs and how to control them.
   d. The crew leader typically prepares the company's cost estimate and is therefore responsible for cost performance.

10. Which of the following is a correct statement regarding project cost?
    a. Cost is handled by the accounting department and is not a concern of the crew leader.
    b. A company's profit on a project is affected by the difference between the estimated cost and the actual cost.
    c. Wasted material is factored into the estimate, and is therefore not a concern.
    d. The contractor's overhead costs are not included in the cost estimate.

11. The crew leader is responsible for ensuring that equipment used by his or her crew is properly maintained.

   a. True
   b. False

12. Who is responsible if a defect in an employee's tool results in an accident?

   a. The employee
   b. The company
   c. The crew leader
   d. The tool manufacturer

13. Productivity is defined as the amount of work accomplished.

   a. True
   b. False

14. If a crew of masons is needed to lay 1,000 concrete blocks, and the estimator determined that two masons could complete the job in one eight-hour day, what is the estimated productivity rate?

   a. 125 blocks per hour
   b. 62.5 blocks per hour
   c. 31.25 blocks per hour
   d. 16 blocks per hour

Additional Resources

This module presents thorough resources for task training. The following resource material is suggested for further study.

Aging Workforce News, www.agingworkforce-news.com.

American Society for Training and Development (ASTD), www.astd.org.

Architecture, Engineering, and Construction Industry (AEC), www.aecinfo.com.

CIT Group, www.citgroup.com.

Equal Employment Opportunity Commission (EEOC), www.eeoc.gov.

National Association of Women in Construction (NAWIC), www.nawic.org.

National Census of Fatal Occupational Injuries (NCFOI), www.bls.gov.

National Center for Construction Education and Research, www.nccer.org.

National Institute of Occupational Safety and Health (NIOSH), www.cdc.gov/niosh.

National Safety Council, www.nsc.org.

NCCER Publications:
Your Role in the Green Environment
Sustainable Construction Supervisor

Occupational Safety and Health Administration (OSHA), www.osha.gov.

Society for Human Resources Management (SHRM), www.shrm.org.

United States Census Bureau, www.census.gov.

United States Department of Labor, www.dol.gov.

USA Today, www.usatoday.com.

Figure Credits

CONTREN® LEARNING SERIES — USER UPDATE

NCCER makes every effort to keep its textbooks up-to-date and free of technical errors. We appreciate your help in this process. If you find an error, a typographical mistake, or an inaccuracy in NCCER's Contren® materials, please fill out this form (or a photocopy), or complete the online form at www.nccer.org/olf. Be sure to include the exact module number, page number, a detailed description, and your recommended correction. Your input will be brought to the attention of the Authoring Team. Thank you for your assistance.

Instructors – If you have an idea for improving this textbook, or have found that additional materials were necessary to teach this module effectively, please let us know so that we may present your suggestions to the Authoring Team.

NCCER Product Development and Revision

3600 NW 43rd Street, Building G, Gainesville, FL 32606

Fax: 352-334-0932
Email: curriculum@nccer.org
Online: www.nccer.org/olf

☐ Trainee Guide ☐ AIG ☐ Exam ☐ PowerPoints Other _____

Craft / Level: _____ Copyright Date: _____

Module Number / Title: _____

Section Number(s): _____

Description: _____

_____

_____

_____

_____

Recommended Correction: _____

_____

_____

_____

Your Name: _____

Address: _____

_____

Email: _____ Phone: _____

Index